Influences of Microplastics on Environmental and Human Health

Influence of Microplastics on Environmental and Human Health

Key Considerations and Future Perspectives

Edited by Yvonne Lang

CRC Press
Taylor & Francis Group
Boca Raton London New York

CRC Press is an imprint of the
Taylor & Francis Group, an **informa** business

First edition published 2022
by CRC Press
6000 Broken Sound Parkway NW, Suite 300, Boca Raton, FL 33487-2742

and by CRC Press
4 Park Square, Milton Park, Abingdon, Oxon, OX14 4RN

CRC Press is an imprint of Taylor & Francis Group, LLC

Library of Congress Cataloging-in-Publication Data
Names: Lang, Yvonne, PhD, editor.
Title: Influence of microplastics on environmental and human health : key considerations and future perspectives / edited by Yvonne Lang.
Description: First edition. | Boca Raton, FL : CRC Press, 2022. | Includes bibliographical references and index.
Identifiers: LCCN 2021053375 | ISBN 9780367612177 (hbk) | ISBN 9780367625702 (pbk) | ISBN 9781003109730 (ebk)
Subjects: LCSH: Microplastics--Environmental aspects. | Plastic scrap--Environmental aspects. | Microplastics--Health aspects. | Microplastics--Toxicology.
Classification: LCC TD427.P62 I54 2022 | DDC 363.738--dc23/eng/20211227
LC record available at https://lccn.loc.gov/2021053375

ISBN: 978-0-367-61217-7 (hbk)
ISBN: 978-0-367-62570-2 (pbk)
ISBN: 978-1-003-10973-0 (ebk)

DOI: 10.1201/9781003109730

Typeset in Minion
by MPS Limited, Dehradun

Contents

Foreword

We live in extraordinary times, and one has only to look around to see all the materials made from plastic that play a central role in our lives, from the clothes we wear to the everyday items we use at home and in the workplace. This has been further conveyed to us during the COVID-19 pandemic and the significance of personal protective equipment (PPE) and single-use sanitary wipes, which are largely made from plastics, and their role in saving lives and protecting health. Initially marketed as miraculous materials owing to their advantageous properties, the success of plastics is reflected by their ever-growing yearly production. While only present on the Earth for little over 100 years, plastic waste is now a geographically widespread problem and has been recorded from the ocean depths to the mountain tops and in many species, including our own. While our comprehension and awareness of microplastics in the environment have intensified over the last decade, knowledge gaps remain. I welcome this new book which provides an important scientific basis for the consideration of microplastics in terms of environmental and human health. From a scientific perspective, while the number of peer-reviewed publications on microplastics in all realms of the environment has increased exponentially in recent years, there is a lack of standardization in methodological approaches for the determination of microplastics in the environment which confounds comparison between studies. The authors identify the main challenge associated with microplastic assessment in terms of analytical approaches from the field to the lab and reinforce the necessity for harmonization in methodologies in order to establish the basis for future perspectives. The impacts of microplastics on human health are considered and the frequently contradictory nature of scientific findings in this area are highlighted. It is clear a better understanding of the bioavailability, biomagnification, and potential toxic effects of exposure to microplastics is required, including from additive compounds used in the production of plastics and both organic and inorganic contaminants adsorbed on microplastic particles and fibres in the natural environment. Owing to their size range and heterogeneity of their properties, microplastics can effectively penetrate through food webs where absorption and desorption of pollutants and inherent chemicals can occur, creating a complex range of potential hazards for biota and humans. In recent years, with the development of more advanced analytical techniques, numerous pollutants of emerging concern have been identified in all compartments of our environment. The chapter on the role of pharmaceutical products on the environmental impact of microplastics is timely as understanding the sources of microplastics in the environment is very important in terms of effective waste management strategies and policy aimed at reducing the global plastics problem. The low settling velocity of MPs and their resistance to degradation means they can be transported long distances through watercourses, making them an emerging pollutant of environmental and human health concern. This book provides a comprehensive overview of both the environmental and human health consequences of microplastics and will be a suitable resource for undergraduate and postgraduate students, teachers, practitioners, and policy makers.

Dr. Liam Morrison
National University of Ireland, Galway

Short Commentary

Challenges and Opportunities for Plastic Waste Management Arising from the COVID-19 Pandemic

Neil J Rowan

Bioscience Research Institute, Athlone Institute of Technology, Ireland

Environmental concerns arising from plastic pollution have created challenges for society to discover alternative and sustainable solutions globally (Silva, 2021). The COVID-19 pandemic caused a rare "black swan" global emergency leading to an unprecedented surge in single-use plastic (SUPs) items such as Personal and Protective Equipment (PPE), along with critical shortages in their supply chains (Klemes et al., 2020; Rowan and Laffey, 2020; Rowan and Moral, 2021). This pandemic also highlighted clear gaps in critical information underpinning efficacy of municipal solid waste (MSW) and disposal systems for dealing with hazardous biomedical waste (Kulkarni and Anantharama, 2020). The volume of domestic or residential waste has also increased by as much as 30% in developed countries due in part to imposed lockdowns (SWANA, 2020). There is a significant disparity between MSW generated between countries, such as Japan and Sweden where >50% is incinerated, ca. 17% is recycled, and ca. 3% is disposed in landfills (Mollicia and Blestieri, 2020), compared with the US and China where a major proportion of MSW is disposed in landfill sites. MSW management is important for ensuring effective public health services, particularly to respond to increased pressures placed on these by SUP activities associated with the COVID-19 pandemic (Thompson, 2020).

Factors considered to influence severe acute respiratory syndrome coronavirus-2 (SARS-CoV2) transmission via MSW handling include: (1) viral persistence or duration of survival on contact surfaces (Rowan and Laffey, 2020); (2) population density that increases usage of SUPs and enhances likelihood of viral transmission (Saadt et al., 2020); and (3) socio-economic conditions where the pandemic is detrimental to people living in poverty (Kulkarni and Anatharam, 2020). Putative implications of the ongoing COVID-19 pandemic on the efficacy of MSW management systems include: (1) intensification of MSW services due to dealing with increased volumes of household and biomedical waste; (2) increased need for recycling that creates challenges for developing countries with relatively low economic status where regulatory levels for treating MSW are under-developed; (3) need for different approaches for sustainable MSW management including thermal treatment (waste to energy) that offers the advantage of hygienisation where hazardous waste and substances are decomposed at very high temperatures (>850°C), mineralized, and immobilized and includes provision for resource recovery (Kulkarni and Anantharama, 2020); and (4) improving or establishing disaster waste management plants. However, it is appreciated that disaster waste management is challenging, expensive, time consuming, and requires broad stakeholder engagement (Rowan and Laffey, 2020). Kulkarni and Anantharama (2020) reported that to ensure uninterrupted MSW management services, and to safeguard personnel involved during COVID-19, the role of automation needs to be explored and solutions applied.

Fan et al. (2021) offered several outlooks based on a review of how the COVID-19 pandemic influences waste management including the need for a longer-term systematic assessment to understand the lasting impact of waste management systems on environmental sustainability and impact. There is a greater need for more transnational and regional studies including provision for modelling to support predictive and simulative solutions to inform decision making (Rowan and Moral, 2021). A more flexible design and planning approach could overcome the restrictive rigidness of waste management systems to include mobile, decentralized, and adaptable (such as exploiting a variety of feedstocks) treatment approaches. Sustaining disruptive development in the adjacent digital domain will also advance waste management including end-to-end monitoring, internet of things, and artificial intelligence including machine learning (Rowan and Casey, 2021). Specialist training and support for managers overseeing MSW including plastic recovery and recycling are merited.

In terms of future-proofing plastic waste management, discoveries and developments in biotechnology linked with polymer science have provided important carbon-neutral building blocks from a plethora of feedstocks encompassing biomass, waste, and residues, along with generating a diverse range of bio-based plastics that have enhanced functionalities and characteristics (Silva, 2021). Silva (2021) highlighted that the need to support and enable a united transition to "circularity" requires creative, if not disruptive, rethinking across the full plastic value chain; therefore, essentially decoupling bio-based plastics from fossil fuels where the latter currently provides 99% of our plastics. Most SUPs are resistant to degradation, which has deleterious environmental consequences as this leads to littering and pollution of natural environments (Beaumont et al., 2019). Global plastic waste production reached 368 million metric tons (Mt) in 2019 that generated ca. 29.1 Mt of plastic waste along with ca. 17.8 Mt for short-life products (such as SUPs) (Silva, 2021); a significant share of this plastic waste (ca. 42%) was inefficiently treated. There is an increasing trend to develop and exploit bio-based plastics, which are plastics based on renewable feedstocks that are considered carbon-neutral alternatives to conventional plastics derived from fossil fuels. Silva (2021) reported on the importance of the "cradle-to-grave life cycle assessment" for bio-based products including harnessing their properties to support reduction in greenhouse gas (GHG) emissions. Silva (2021) and others (Rowan and Casey, 2021; Galanakis et al., 2021) reported on the biorefinery agricultural, forest, and food waste streams for lipids, flavoids, cellulose, lignin, and phenolic compounds as added value bio-based products. There is also increasing focus on algae-based biopolymers as future feedstocks (including plants based) that include advantages of cultivation costs, non-competition with land-based feedstocks, and autotrophy that reduce GHG emissions (Rowan and Galanakis, 2020). There are also greater opportunities for sustainable waste management using bio-based plastics that promote "circularity." However, advances in biotechnology are required to accelerate this ambition that will be achieved in part by investing in public education in order to inform behavioural change, acceptance, and awareness for bio-based, biodegradable (e.g. PHA), compostable, and recycled products. Silva (2021) noted that several drop-in bio-based plastics are already commercially available that undergo similar end-of-life modalities, such as bioPE and bioPET. Moreover, these bio-based products are accommodated by established recycling processes due to similarity in mechanical and chemical characteristics to that of PE and PET (Silva, 2021).

In summary, there is a pressing need to improve plastic waste management including the development of alternative bio-based products that should be suitable for reuse, recycling, and energy recovery. There is a commensurate need for regulations and policies that encourage manufacturers to develop safe-by-design products along with deploying life cycle assessment and social marketing to inform behavioural change for end-users. Several researchers have also noted the importance of considering valorizing bio-based plastic products, wastes, and residues along with composting and anaerobic digestion (Silva, 2021).

REFERENCES

Beaumont, N. J., Mouneyrac, M., Costa, M., Duarte, A. C. & Rocha-Santos, R. (2020) The role of legislation, regulatory initiatives and guidelines on the control of plastic pollution. *Frontiers in Environmental Science*, 8, 104. https://doi.org/103389/fenvs.2020.00104.

Fan, Y. V., Jiang, P., Hemzal, M. & Klemes, J. J. (2021) An update of COVID-19 influence on waste management. *Science of the Total Environment*. https://doi.org/10.1016/j.scitotenv.2020.142014.

Galanakis, C. M., Rizou, M., Aldawoud, T. M. S., Ucak, I. & Rowan, N. J. (2021) Innovations and technology disruptions in the food sector within the COVID-19 pandemic and post-lockdown era. *Trends in Food Science and Technology*, 110, 193–200.

Klemes, J. J., Fan, Y. V., Tan, R. R. & Jiang, P. (2020) Minimizing the present and future plastic waste, energy and environmental footprints related to COVID-19. *Renewable and Sustainable Energy Reviews*, 127, 109883. https://doi.org/101016/jrser.2020.109883.

Kulkarni, B. N. & Anantharama, V. (2020) Repercussions of COVID-19 pandemic on municipal solid waste management: Challenges and opportunities. *Science of the Total Environment*. https://doi.org/10.1016/j.scitotenv.2020.140693.

Mollica, G. J. G. & Balestierai, J. A. P. (2020) Is it worth generating energy with garbage? Defining a carbon tax to encourage waste-to-energy cycles. *Applied Thermal Engineering*, 173. https://doi.org/10.1016/j.applthermaleng.2020.115195.

Rowan, N. J. & Moral, R. A. (2021) Disposable face masks and reusable face coverings as non-pharmaceutical interventions (NPIs) to prevent transmission of SARS-CoV-2 variants that cause coronavirus disease (COVID-19): Role of new sustainable NPI design innovations and predictive mathematical modelling. *Science of the Total Environment*. https://doi.org/10.1016/j.scitotenv.2021.145530.

Rowan, N. J. & Casey. O. (2021) Empower Eco multiactor HUB: A triple helix 'academia-industry-authority' approach to creating and sharing potentially disruptive tools for addressing novel and emerging new

Green Deal opportunities under a United Nations Sustainable Development Goals framework. *Current Opinion in Environmental Science & Health*, 21. https://doi.org/10.1016/j.coesh.2021.100254.

Rowan, N. J. & Galankis, C. M. (2020) Unlocking challenges and opportunities presented by COVID-19 pandemic for cross-cutting disruption in agri-food and green deal innovations: Quo Vadis? *Science of the Total Environment.* https://doi.org/10.1016/j.scitotenv.2020.141362.

Rowan, N. J. & Laffey, J. G. (2020) Unlocking the surge in demand for personal and protective equipment (PPE) and improvised face coverings arising from coronavirus disease (COVID-19) pandemic implications for efficacy, reuse and sustainable waste management. *Science of the Total Environment.* https://doi.org/10.1016/j.scitotenv.2020.142259.

Saadat, S., Rawtani, D. & Hussain, C. M. (2020) Environmental perspective of COVID-19. *Science of the Total Environment*, 738. https://doi.org/10.1016/j.scitotenv.2020.138870.

Silva, A. L. P. (2021) Future-proofing plastic waste management for a circular bio-economy. *Current Opinion in Environmental Science and Health.* https://doi.org/10.1016/j.coesh.2021.100263.

SWANA–Solid Waste Association of North America article (2020) SWANA reminds all state and local governments that solid waste management is an essential public service. https://swan.org/news/swan-news/article/202003/19/swan-reminds-all-state-and-local-governments-that-solid-waste-management-is-an-essential-public-service (accessed 24 June, 2021).

Thompson, B. (2020) The COVID-19 pandemic: A global natural experiment. *Circulation.* https://doi.org/10.1161/CIRCULATIONAHA.120.047538.

Contributors

João Frias
Marine and Freshwater
  Research Centre (MFRC)
Galway-Mayo Institute of
  Technology (GMIT)
Galway, Ireland

Mary Heneghan
Fungal Molecular Biology
  Group
Institute of Technology Sligo
Sligo, Ireland
and
Centre for Environmental
  Research, Innovation and
  Sustainability (CERIS)
Institute of Technology Sligo
Sligo, Ireland

Naveen Kumar
Fungal Molecular Biology
  Group
Institute of Technology Sligo
Sligo, Ireland

Yvonne Lang
Centre for Environmental
  Research, Innovation and
  Sustainability (CERIS)
Institute of Technology Sligo
Sligo, Ireland

Jenny Lawler
School of Biotechnology and
  DCU Water Institute
Dublin City University
DCU Glasnevin Campus
Dublin, Ireland
Qatar Environment and
  Energy Research Institute
Hamad Bin Khalifa University
Doha, Qatar

Róisín Nash
Marine and Freshwater
  Research Centre (MFRC)
Galway-Mayo Institute of
  Technology (GMIT)
Galway, Ireland

Irene O'Callaghan
School of Biological, Earth and
    Environmental Sciences
University College Cork
Ireland
School of Chemical Sciences
University College Cork
Ireland

Sandra O'Neill
School of Biotechnology and
    DCU Water Institute
Dublin City University
DCU Glasnevin Campus
Dublin, Ireland

Suresh C. Pillai
Nanotechnology &
    Bioengineering Research
    Group (Nano-Bio)
Institute of Technology Sligo
Sligo, Ireland

Timothy Sullivan
School of Biological, Earth and
    Environmental Sciences
University College Cork
Ireland
Environmental Research
    Institute
University College Cork
Ireland

CHAPTER 1

How Do Nanoplastics and Microplastics Impact Human Health?

Yvonne Lang

Centre for Environmental Research, Innovation and Sustainability (CERIS), Institute of Technology Sligo, Ash Lane, SligoF91 YW50, Ireland

CONTENTS

DOI: 10.1201/9781003109730-1

1.1 PLASTIC POLLUTION AND THE "WAR ON PLASTIC"

Nanoplastics (NPs) and microplastics (MPs) are plastic particles that measure less than 5 mm in any one dimension (Arthur et al, 2009; Thompson et al, 2004). The research community and policymakers have acknowledged that the scale of plastic pollution (including NPs and MPs) presents an immediate threat to environmental health and human health. Specific tasks, targets, and actions to address plastic pollution are included in The UN Sustainable Development Goals (Walker, 2021). The European Union Commission released the Strategy on Plastics in 2018 (European Commission, 2018) and the Directive on Single-Use Plastics in 2019 (European Commission, 2019) to provide a starting point to reduce plastic pollution and begin the journey towards a circular plastic economy. The result of these actions will have a positive influence in tackling NP and MP pollution.

Key reports published in 2019 and 2020 by leading scientific experts summarized current knowledge on environmental and human health risks of NPs and MPs (European Commission's Scientific Advice Mechanism, 2019; GESAMP, 2020; SAPEA, 2019; World Health Organisation, 2019). Each report reviewed available data in order to discern what was clear evidence of the impacts of NPs and MPs, and what was proposed, suggested, or speculated impacts of NPs and MPs. Overall there was a consensus across reports that due to the immaturity of the research field, and the multifaceted nature of evaluating the impact of NPs and MPs scientific knowledge "*is a mix of consensus, contested knowledge, informed extrapolation, speculation, and many unknowns*"(European Commission's Scientific Advice Mechanism, 2019). Currently, the literature does not provide clear and unambiguous evidence that NPs and/or MPs have an adverse effect on human health. However, it is strongly recommended that action is taken to prevent further release of these particles into the environment and to reduce and/or remove potential human health hazards.

Points raised in the reports with respect to challenges that are faced in assessing the impact of NPs and/or MPs on human health include

- lack of harmonization in the methods of assessment of exposure to NPs and MPs

- the potential that laboratory experiments do not relate to real-world exposure

- the impact of exposure to NPs and MPs on human health is influenced by the physicochemical properties of the particles, but also potential interactions of the NPs and MPs with other substances be it chemical or biologically when released into the environment

The Joint Group of Experts on the Scientific Aspects of Marine Environmental Protection identified seven key challenges to address risk assessment ranging from pathways of exposure, quantification of exposure, development of methods and technologies for monitoring and assessing exposure, upscaling effects, communication of the uncertainties and limitations, and integrating risk assessment into the policy decision process (GESAMP, 2020).

The challenges and recommendations presented by the various expert groups provide a roadmap for future directions of research in the field of NPs and/or MPs. It is time that such direction is provided as it is acknowledged that there has been an exponential growth of publications in the field of NPs and MPs. Figure 1.1 illustrates the results of a search of the Web of Science database using the keyword searches (1) "Microplastic" OR "Nanoplastic" AND "Human Health" (2) "Microplastic" OR "Nanoplastic" AND "Environment" within the period 2000–2020. Search results for January to September 2021 generate 1081 hits and 1107 hits respectively. This demonstrates the volume of work being conducted in the context of environmental and health impacts of NPs and

TABLE 1.1 *In vitro* viability studies investigating effects of exposure to NPs or MPs

Particle size	Concentration range(s)	NP/MP	Exposure time (hours)	Cell line (s)	Reference
100 nm	5–75 µg/mL	PS	4–48	Human Fibroblast Hs27 cells	(Poma et al, 2019)
60 nm	1–10 µg/mL	PS	24	Human breast adenocarcinoma cell lines MDA-MB 231 and MCF-7	(Roje et al, 2019)
50 nm 0.5 µm	0.1–100 µg/mL	PS	24	Caco-2 and HT29-MTX-E12 co-cultures	(Hesler et al, 2019)
0.1 µm 5 µm	1–200 µg/mL	PS	12	Caco-2 cells	(Wu et al, 2019)
25 nm 70 nm	2.5–30 µg/mL 10–300 µg/mL	PS	24	A549 human lung epithelial cells	(Xu et al, 2019)
100 nm	1–25 µg/mL	PS	24	Lymphocytes	(Gopinath et al, 2019)
300 nm 500 nm 6 µm	Not stated	PS	24	Caco-2 cells	(Wang et al, 2020)
460 nm 1 µm 3 µm	1–1000 µg/mL	PS	Not stated	Human dermal fibroblast cells (HDF) Human peripheral blood mononuclear cells (PBMC)	(Hwang et al, 2020)

Size	Concentration	Polymer	Time (h)	Cell model	Reference
10 μm					
40 μm					
100 μm					
1.67–2.17 μm	1–1000 μg/cm^2	PS	24–48	BEAS-2B human lung epithelial cells	(Dong et al, 2020)
50 nm	10–100 μg/mL	PS	24	Human hepatocellular carcinoma (HepG2) cells	(He et al, 2020)
1–4 μm 10–20 μm	100 mg/ml	PE, PP, PET, PVC	24	Caco-2 cells, HepG2 and HepaRG	(Stock et al, 2021)
50 nm < 50 μm	1, 5, 10 or 50 μg/cm^2	PS and PVC	24	Caco-2/HT29-MTX-E12/THP-1 triple culture model	(Busch et al, 2021)

PS, polystyrene; PP, polypropylene; PET, polyethylene terephthalate; PVC, poylvinyl chloride.

cytotoxic (Dong et al, 2020; Gopinath et al, 2019), genotoxic (Gopinath et al, 2019), embryotoxic (Hesler et al, 2019), inflammatory (Dong et al, 2020), and immunomodulatory effects (Hwang et al, 2019) and are in agreement with previous knowledge.

It must be questioned whether this vast amount of data provides conclusive evidence of harm to human health considering the limitations and difficulties of comparing studies. This is evidenced in the concluding statements of some of the studies, where there is an acknowledgement that the results are suggestive of possible harm, and provide useful information for areas of emerging interest, but ultimately further studies are necessary (Dong et al, 2020; Gopinath et al, 2019; Xu et al, 2019).

In order to advance knowledge *in vitro* studies need to begin assessing NPs and/or MPs in a manner that is reflective of the real-world situation. This includes consideration of the following when designing the *in vitro* study:

- Is evaluation of pristine or virgin plastics advancing knowledge of impacts on human health?

- Is the examination of spherical particles solely a limitation as NPs and/or MPs are formed by degradation of larger plastics and fragments will have neither a consistent size nor shape?

- Studies examining the influence of size need to ensure that the number of particles is normalized across the different size particles. For example, is it correct to compare exposure of cells to a 20 µg/ml solution of 0.1 µm PS spheres to 20 µg/ml solution of 5 µm PS spheres? The smaller spheres will be present in a higher number despite the weight equivalence. Also, is there potential for aggregation? Aggregation of NPs and/or MPs will have a bearing on the number of particles that the body is exposed and thus study design needs to advance to evaluate monodispersed systems and aggregated systems.

- It is important to provide evidence of the cell:particle ratio in order to compare study outcomes of cell lines. For example, is it correct to compare the outcomes of studies conducted in a Caco-2 cell line where one has a seeded cell density of 1.5×10^4 cells/ml and the other has a seeded cell density of 5×10^4 cells/ml?

- Studies need to clearly distinguish between outcomes that are statistically relevant but not biologically relevant and those that are statistically relevant and biologically relevant.

1.2.2 *In Vivo* Studies of NPs and MPs in Rodent Models

There is an acknowledgement by authors of *in vivo* studies that the doses used are difficult to relate to the real world. While MPs have been detected in human faeces (Schwabl et al, 2019), human lung tissue (Amato-Lourenço et al, 2021), and human placenta (Ragusa et al, 2021), there is still a lack of knowledge quantifying the extent of exposure to NPs/MPs that humans will experience. Various estimates of the possible number of NPs and/or MPs that humans may intake via different routes of exposure are reported (Domenech & Marcos, 2021; Mohamed Nor et al, 2021) with no consensus as yet. Thus, *in vivo* rodent studies aim to demonstrate the impact of exposure at various levels. As the actual exposure levels become more apparent it will be possible to relate the observations from *in vivo* studies to potential risk. Also, there is an acknowledgement that the use of NP and MP preparations that have not undergone weathering and are typically spherical in nature may not be reflective of the real-world environment. Thus the conclusion of many animal studies is that the data points to a potential risk but it is not a declaration of fact. Table 1.2 summarizes the key findings of *in vivo* studies in rodent models published since the SAPEA 2019 report. Studies have reported alterations to the gastrointestinal system

TABLE 1.2 *In vivo* effects of exposure to nanoplastics or microplastics

NP/MP	NP/MP exposure	Key findings	Reference
PS	Exposure in drinking water at 100 and 1000 µg/L for five weeks	Alterations to the gut microbiota	(Lu et al, 2018)
PS	Exposure in drinking water at 100 and 1000 µg/L for gestation	Alterations indicative of metabolic disorders in offspring	(Luo et al, 2019)
PS	Exposure in drinking water at 100 and 1000 µg/L for six weeks	Alterations to the gut microbiota and integrity of the intestinal membrane	(Jin et al, 2019)
PS	Exposure at 1, 3, 6, and 10 mg/kg-day for five weeks	Abnormal sperm quality	(Amereh et al, 2020)
PS	Exposure at 0.01 mg/day, 0.1 mg/day, and 1 mg/day for 42 days	Abnormal sperm quality	(Xie et al, 2020)
PE	Exposure at 6, 60, and 600 mg/day for 5 consecutive weeks	Alterations to the gut microbiota and inflammation of the intestine	(Li et al, 2020)
PE	Exposure at 0.125 mg/day, 0.5 mg/day, and 2 mg/day for 90 days	Alterations in the health of offspring	(Park et al, 2020)
PS	Exposure at 0.6–0.7 µg/day, 6–7 µg/day, and 60–70 µg/day for 35 days	Abnormal sperm quality	(Hou et al, 2021)
PS	Exposure at 0.015, 0.15 and 1.5 mg/day for 90 days	Alterations to ovarian reserve capacity	(An et al, 2021)

PS, polystyrene; PE, polyethylene.

(Jin et al, 2019; Li et al, 2020; Lu et al, 2018), reproductive systems (Amereh et al, 2020; An et al, 2021; Hou et al, 2021; Xie et al, 2020), and progeny (Luo et al, 2019; Park et al, 2020).

Similar to the scenario presented regarding *in vitro* data, these results do not provide conclusive evidence of harm to human health. However, it does provide sufficient concern to continue with a precautionary approach to human exposure to NPs and MPs as inaction may lead to detrimental health effects in future years. (Leslie & Depledge, 2020) posed a very pertinent question *"Can the conclusion of 'no risk' be supported by 'no data'? One of the common pitfalls in critical thinking is to neglect the logic that the absence of evidence is not evidence of absence."*

1.3 FUTURE CONSIDERATIONS

There have been a number of interesting reviews published on specific elements of potential human health impacts of NPs/MPs since the 2019 SAPEA report was released including the intestinal environment and intestinal barrier (Huang et al, 2021), the digestive, reproductive, and nervous systems (Yin et al, 2021), and antibiotic resistance (Liu et al, 2021; Pham et al, 2021). However, human health impacts are not merely due to the direct exposure to NPs and/or MPs. Detrimental effects of NPs and/or MPs on marine, freshwater, and terrestrial environments will have an impact on human health. The most obvious one of these is the deterioration of ocean health and soil health due to the presence of NPs and/or MPs. Understanding the scale and complexity of consequences to human health due to this deterioration is going to require multidisciplinary approaches and collaborations. It is outside the scope of this chapter to discuss this in detail and the reader is referred to recent comprehensive reviews exploring relationships between NPs and/or MPs and topics such biodiversity loss (Agathokleous et al, 2021), ecosystem services (Sridharan et al, 2021), the aquatic microbiome (Nava & Leoni, 2021; Yang et al, 2020), vectors for substances of concern in the environment (Bradney et al, 2019; Torres et al, 2021; Vieira et al, 2021), and food safety (Barboza et al, 2018). Undoubtedly, further questions will be posed as new knowledge emerges.

Protection of environmental health lies in the hands of humankind and it is important that the impact of NPs and/or MPs do not overshadow the overarching need to reduce plastic consumption and waste. Rist et al, 2018 comment that by focusing on NPs and/or MPs without due consideration for the broader issues that:

> "We risk pulling the focus away from the root of the problem: the way in which we consume, use and dispose of plastics leading to their widespread presence in our everyday life and in the environment."

Therefore, the importance of the public perception of the plastic issue overall and NPs and MPs needs to be assessed so that enhanced engagement and behavioural changes can occur to assist in addressing the "war on plastic."

Public perception of the risk of NPs and MPs will be influenced by the information source referenced. There is a responsibility on the scientific community to ensure that the real risks and proposed risks are presented in a transparent manner. Work by Völker et al, 2020 examined the manner by which scientific papers present the environmental risk associated with MPs. Four hundred and sixty-four articles published between 2006 and 2018 were examined with 66.8% discussing hypothetical environmental risk, and 24.4% stating an environmental risk exists. An interesting finding of the work by Völker et al, 2020 was that the analysis of 97 media articles revealed that only 7.2% were classified as a "factual representation of scientific findings without further interpretations or the use of judgmental terms." Furthermore, of the 97 articles, 47.4% communicated scientific hypotheses about the impacts of MP pollution as scientific facts.

The role of the media and scientific community in presenting information on MPs in a coherent, complimentary, and clear manner is paramount to ensuring that misinformation and exaggerated scenarios of the impact of NPs and/or MPs on human health are avoided (Thiele & Hudson, 2021).

The intensity of research to evaluate the impacts of NPs and MPs on both environmental health and human health is representative of the desire for change by researchers, policymakers, and the general public. It is critical that new knowledge is generated as a result of these efforts and research must address gaps in knowledge. Bonanno & Orlando-Bonaca, 2018 presented important points to consider as we move forward designing policy and identifying where research funding should be invested. These questions range from realistic estimates of plastics in the sea, sources of plastics, the fate of the plastics released into the sea, ecological effects of plastics in the sea, transfer of plastic across trophic levels, and marine plastic impact on human health.

It has been conclusively demonstrated that NPs and MPs have had detrimental effects on environmental health. As human health is inherently linked to environmental health, it can be speculated that it is inevitable human health will be negatively impacted by NPs and MPs, and it is only a matter of time before this manifests itself.

REFERENCES

Agathokleous, E., Iavicoli, I., Barceló, D. & Calabrese, E. J. (2021) Ecological risks in a 'plastic' world: A threat to biological diversity? *Journal of Hazardous Materials*, 417, 126035.

Amato-Lourenço, L. F., Carvalho-Oliveira, R., Júnior, G. R., dos Santos Galvão, L., Ando, R. A. & Mauad, T. (2021) Presence of airborne microplastics in human lung tissue. *Journal of Hazardous Materials*, 416, 126124.

Amereh, F., Babaei, M., Eslami, A., Fazelipour, S. & Rafiee, M. (2020) The emerging risk of exposure to nano(micro)plastics on endocrine disturbance and reproductive toxicity: From a hypothetical scenario to a global public health challenge. *Environmental Pollution*, 261, 114158.

An, R., Wang, X., Yang, L., Zhang, J., Wang, N., Xu, F., Hou, Y., Zhang, H. & Zhang, L. (2021) Polystyrene microplastics cause granulosa cells apoptosis and fibrosis in ovary through oxidative stress in rats. *Toxicology*, 449, 152665.

Arthur, C., Baker, J., Bamford, H. (eds) (2009) Proceedings of the International Research Workshop on the Occurrence, Effects and Fate of Microplastic Marine Debris., *NOAA Technical Memorandum NOS-OR&R-30*.

Barboza, L. G. A., Dick Vethaak, A., Lavorante, B. R. B. O., Lundebye, A.-K. & Guilhermino, L. (2018) Marine microplastic debris: An emerging issue for food security, food safety and human health. *Marine Pollution Bulletin*, 133, 336–348.

Bonanno, G. & Orlando-Bonaca, M. (2018) Ten inconvenient questions about plastics in the sea. *Environmental Science & Policy*, 85, 146–154.

Bradney, L., Wijesekara, H., Palansooriya, K. N., Obadamudalige, N., Bolan, N. S., Ok, Y. S., Rinklebe, J., Kim, K.-H. & Kirkham, M. B. (2019) Particulate plastics as a vector for toxic trace-element uptake by aquatic and terrestrial organisms and human health risk. *Environment International*, 131, 104937.

Busch, M., Bredeck, G., Kampfer, A. A. M. & Schins, R. P. F. (2021) Investigations of acute effects of polystyrene and polyvinyl chloride micro- and nanoplastics in an advanced in vitro triple culture model of the healthy and inflamed intestine. *Environmental Research*, 193.

Campanale, C., Massarelli, C., Savino, I., Locaputo, V. & Uricchio, V. F. (2020) A detailed review study on potential effects of microplastics and additives of concern on human health. *International Journal of Environmental Research and Public Health*, 17(4), 1212.

Domenech, J. & Marcos, R. (2021) Pathways of human exposure to microplastics, and estimation of the total burden. *Current Opinion in Food Science*, 39, 144–151.

Dong, C. D., Chen, C. W., Chen, Y. C., Chen, H. H., Lee, J. S. & Lin, C. H. (2020) Polystyrene microplastic particles: In vitro pulmonary toxicity assessment. *Journal of Hazardous Materials*, 385.

European Commission (2018) COMMUNICATION FROM THE COMMISSION TO THE EUROPEAN PARLIAMENT, THE COUNCIL, THE EUROPEAN ECONOMIC AND SOCIAL COMMITTEE AND THE COMMITTEE OF THE REGIONS A European Strategy for Plastics in a Circular Economy COM/2018/ 028 final.

European Commission (2019) Directive (EU) 2019/904 of the European Parliament and of the Council of 5 June 2019 on the reduction of the impact of certain plastic products on the environment PE/11/2019/REV/1

European Commission's Scientific Advice Mechanism (2019) *Environmental and health risks of microplastic pollution.* Brussels.

Garcia-Vazquez, E., Garcia-Ael, C. & Topa, G. (2021) On the way to reduce marine microplastics pollution. Research landscape of psychosocial drivers. *Science of The Total Environment*, 799, 149384.

GESAMP (2020) *Proceedings of the GESAMP International Workshop on assessing the risks associated with plastics and microplastics in the marine environment.*

Gonzalez-Acedo, A., Garcia-Recio, E., Illescas-Montes, R., Ramos-Torrecillas, J., Melguizo-Rodriguez, L. & Costela-Ruiz, V. J. (2021) Evidence from in vitro and in vivo studies on the potential health repercussions of micro- and nanoplastics. *Chemosphere*, 280, 130826.

Gopinath, P. M., Saranya, V., Vijayakumar, S., Meera, M. M., Ruprekha, S., Kunal, R., Pranay, A., Thomas, J., Mukherjee, A. & Chandrasekaran, N. (2019) Assessment on interactive prospectives of nanoplastics with plasma proteins and the toxicological impacts of virgin, coronated and environmentally released-nanoplastics. *Scientific Reports*, 9.

He, Y. X., Li, J., Chen, J. C., Miao, X. J., Li, G., He, Q., Xu, H. Z., Li, H. & Wei, Y. Y. (2020) Cytotoxic effects of polystyrene nanoplastics with different surface functionalization on human HepG2 cells. *Science of the Total Environment*, 723, 138180.

Hesler, M., Aengenheister, L., Ellinger, B., Drexel, R., Straskraba, S., Jost, C., Wagner, S., Meier, F., von Briesen, H., Büchel, C., Wick, P., Buerki-Thurnherr, T. & Kohl, Y. (2019) Multi-endpoint toxicological assessment of polystyrene nano- and microparticles in different biological models in vitro. *Toxicology in Vitro*, 61, 104610.

Hou, B., Wang, F., Liu, T. & Wang, Z. (2021) Reproductive toxicity of polystyrene microplastics: In vivo experimental study on testicular toxicity in mice. *Journal of Hazardous Materials*, 405, 124028.

Huang, Z., Weng, Y., Shen, Q., Zhao, Y. & Jin, Y. (2021) Microplastic: A potential threat to human and animal health by interfering with the intestinal barrier function and changing the intestinal microenvironment. *Science of The Total Environment*, 785, 147365.

Hwang, J., Choi, D., Han, S., Choi, J. & Hong, J. (2019) An assessment of the toxicity of polypropylene microplastics in human derived cells. *Science of the Total Environment*, 684, 657–669.

Hwang, J., Choi, D., Han, S., Jung, S., Choi, J. & Hong, J. (2020) Potential toxicity of polystyrene microplastic particles. *Scientific Reports*, 10(1).

Jin, Y., Lu, L., Tu, W., Luo, T. & Fu, Z. (2019) Impacts of polystyrene microplastic on the gut barrier, microbiota and metabolism of mice. *Science of The Total Environment*, 649, 308–317.

Lehner, R., Weder, C., Petri-Fink, A. & Rothen-Rutishauser, B. (2019) Emergence of nanoplastic in the environment and possible impact on human health. *Environmental Science & Technology*, 53(4), 1748–1765.

Leslie, H. A. & Depledge, M. H. (2020) Where is the evidence that human exposure to microplastics is safe? *Environment International*, 142, 3.

Li, B., Ding, Y., Cheng, X., Sheng, D., Xu, Z., Rong, Q., Wu, Y., Zhao, H., Ji, X. & Zhang, Y. (2020) Polyethylene microplastics affect the distribution of gut microbiota and inflammation development in mice. *Chemosphere*, 244, 125492.

Liu, Y., Liu, W., Yang, X., Wang, J., Lin, H. & Yang, Y. (2021) Microplastics are a hotspot for antibiotic resistance genes: Progress and perspective. *Science of The Total Environment*, 773, 145643.

Lu, L., Wan, Z., Luo, T., Fu, Z. & Jin, Y. (2018) Polystyrene microplastics induce gut microbiota dysbiosis and hepatic lipid metabolism disorder in mice. *Science of The Total Environment*, 631-632, 449–458.

Luo, T., Zhang, Y., Wang, C., Wang, X., Zhou, J., Shen, M., Zhao, Y., Fu, Z. & Jin, Y. (2019) Maternal exposure to different sizes of polystyrene microplastics during gestation causes metabolic disorders in their offspring. *Environmental Pollution*, 255, 113122.

Mohamed Nor, N. H., Kooi, M., Diepens, N. J. & Koelmans, A. A. (2021) Lifetime accumulation of microplastic in children and adults. *Environmental Science & Technology*, 55(8), 5084–5096.

Nava, V. & Leoni, B. (2021) A critical review of interactions between microplastics, microalgae and aquatic ecosystem function. *Water Research*, 188, 116476.

Park, E.-J., Han, J.-S., Park, E.-J., Seong, E., Lee, G.-H., Kim, D.-W., Son, H.-Y., Han, H.-Y. & Lee, B.-S. (2020) Repeated-oral dose toxicity of polyethylene microplastics and the possible implications on reproduction and development of the next generation. *Toxicology Letters*, 324, 75–85.

Pham, D. N., Clark, L. & Li, M. (2021) Microplastics as hubs enriching antibiotic-resistant bacteria and pathogens in municipal activated sludge. *Journal of Hazardous Materials Letters*, 2, 100014.

Poma, A., Vecchiotti, G., Colafarina, S., Zarivi, O., Aloisi, M., Arrizza, L., Chichiriccò, G. & Di Carlo, P. (2019) In vitro genotoxicity of polystyrene nanoparticles on the human fibroblast Hs27 cell line. *Nanomaterials (Basel, Switzerland)*, 9(9), 1299.

Prata, J. C., da Costa, J. P., Lopes, I., Duarte, A. C. & Rocha-Santos, T. (2020) Environmental exposure to microplastics: An overview on possible human health effects. *Science of the Total Environment*, 702.

Qin, M., Chen, C. Y., Song, B., Shen, M. C., Cao, W. C., Yang, H. L., Zeng, G. M. & Gong, J. L. (2021) A review of biodegradable plastics to biodegradable microplastics: Another ecological threat to soil environments? *Journal of Cleaner Production*, 312.

Ragusa, A., Svelato, A., Santacroce, C., Catalano, P., Notarstefano, V., Carnevali, O., Papa, F., Rongioletti, M. C. A., Baiocco, F., Draghi, S., D'Amore, E., Rinaldo, D., Matta, M. & Giorgini, E. (2021) Plasticenta: First evidence of microplastics in human placenta. *Environment International*, 146.

Rist, S., Carney Almroth, B., Hartmann, N. B. & Karlsson, T. M. (2018) A critical perspective on early communications concerning human health aspects of microplastics. *Science of The Total Environment*, 626, 720–726.

Roje, Ž., Ilić, K., Galić, E., Pavičić, I., Turčić, P., Stanec, Z. & Vrček, I. V. (2019) Synergistic effects of parabens and plastic nanoparticles on proliferation of human breast cancer cells. *Archives of Industrial Hygiene and Toxicology*, 70(4), 310–314.

SAPEA (2019) *A Scientific Perspective on Microplastics in Nature and Society*. Berlin.

Schwabl, P., Koppel, S., Konigshofer, P., Bucsics, T., Trauner, M., Reiberger, T. & Liebmann, B. (2019) Detection of various microplastics in human stool a prospective case series. *Annals of Internal Medicine*, 171(7), 453-+.

Sridharan, S., Kumar, M., Bolan, N. S., Singh, L., Kumar, S., Kumar, R. & You, S. (2021) Are microplastics destabilizing the global network of terrestrial and aquatic ecosystem services? *Environmental Research*, 198, 111243.

Stock, V., Laurisch, C., Franke, J., Donmez, M. H., Voss, L., Bohmert, L., Braeuning, A. & Sieg, H. (2021) Uptake and cellular effects of PE, PP, PET and PVC microplastic particles. *Toxicology in Vitro*, 70.

Thiele, C. J. & Hudson, M. D. (2021) Uncertainty about the risks associated with microplastics among lay and topic-experienced respondents. *Sci Rep*, 11(1), 7155.

Thompson, R. C., Olsen, Y., Mitchell, R. P., Davis, A., Rowland, S. J., John, A. W. G., McGonigle, D. & Russell, A. E. (2004) Lost at sea: Where is all the plastic? *Science*, 304(5672), 838.

Torres, F. G., Dioses-Salinas, D. C., Pizarro-Ortega, C. I. & De-la-Torre, G. E. (2021) Sorption of chemical contaminants on degradable and non-degradable microplastics: Recent progress and research trends. *Science of The Total Environment*, 757, 143875.

Vieira, Y., Lima, E. C., Foletto, E. L. & Dotto, G. L. (2021) Microplastics physicochemical properties, specific adsorption modeling and their interaction with pharmaceuticals and other emerging contaminants. *Science of the Total Environment*, 753, 141981.

Völker, C., Kramm, J. & Wagner, M. (2020) On the creation of risk: Framing of microplastics risks in science and media. *Global Challenges*, 4(6), 1900010.

Walker, T. R. (2021) (Micro)plastics and the UN sustainable development goals. *Current Opinion in Green and Sustainable Chemistry*, 30, 100497.

Wang, Q., Bai, J., Ning, B., Fan, L., Sun, T., Fang, Y., Wu, J., Li, S., Duan, C., Zhang, Y., Liang, J. & Gao, Z. (2020) Effects of bisphenol A and nanoscale and microscale polystyrene plastic exposure on particle uptake and toxicity in human Caco-2 cells. *Chemosphere*, 254, 126788.

World Health Organisation (2019) *Microplastics in drinking-water*. Geneva.

Wu, B., Wu, X., Liu, S., Wang, Z. & Chen, L. (2019) Size-dependent effects of polystyrene microplastics on cytotoxicity and efflux pump inhibition in human Caco-2 cells. *Chemosphere*, 221, 333–341.

Xie, X., Deng, T., Duan, J., Xie, J., Yuan, J. & Chen, M. (2020) Exposure to polystyrene microplastics causes reproductive toxicity through oxidative stress and activation of the p38 MAPK signaling pathway. *Ecotoxicology and Environmental Safety*, 190, 110133.

Xu, M. K., Halimu, G., Zhang, Q. R., Song, Y. B., Fu, X. H., Li, Y. Q., Li, Y. S. & Zhang, H. W. (2019) Internalization and toxicity: A preliminary study of effects of nanoplastic particles on human lung epithelial cell. *Science of the Total Environment*, 694.

Yang, Y., Liu, W., Zhang, Z., Grossart, H.-P. & Gadd, G. M. (2020) Microplastics provide new microbial niches in aquatic environments. *Applied Microbiology and Biotechnology*, 104(15), 6501–6511.

Yin, K., Wang, Y., Zhao, H., Wang, D., Guo, M., Mu, M., Liu, Y., Nie, X., Li, B., Li, J. & Xing, M. (2021) A comparative review of microplastics and nanoplastics: Toxicity hazards on digestive, reproductive and nervous system. *Science of The Total Environment*, 774, 145758.

CHAPTER 2

Methodologies to Assess Microplastics in the Anthropocene

João Frias and Róisín Nash

Marine and Freshwater Research Centre (MFRC), Galway-Mayo Institute of Technology (GMIT), Galway, Ireland

CONTENTS

2.1 GENESIS AND RELEVANCE OF THE PROBLEM

Throughout history, human curiosity led to discoveries (e.g. fire, soap), and new methodologies (e.g. woodwork, metalwork) that enabled the development of our species. The transition from the hunter-gatherer lifestyle into an agriculture-based settlement made humans more dependent on climate and on the finite natural resources surrounding the settlements (Harari, 2015).

DOI: 10.1201/9781003109730-2

These conditions pushed the limits of human creativity to develop or improve tools and methods that would make processes more efficient. Historically those tools were made from the most common material available: wood, stone, bronze, iron, etc. As settlement numbers started to increase and expand in an area, humans had to adapt and improve their systems (Harari, 2015).

Centuries passed and development brought large-scale production of goods and new machinery which led to the Industrial Revolution and to the Machine Age. In this golden age of discoveries, Belgium chemist Leo Baekland would initiate the momentum in polymer science with the discovery of the first synthetic polymer in 1907 (Shashoua, 2008). Supply and demand led to the opening of the first supermarkets between the 1910s and 1930s (Martinho & Cunha, 2010) which contributed to the need for suitable and reliable packaging. Subsequently, after World War II, around the 1950s, worldwide plastic production began (PlasticsEurope, 2010), and in the convening years has grown exponentially at a rate of 8% per annum (PlasticsEurope, 2010 & 2020). By 2020, there were more than 50 different basic types of polymers (PlasticsEurope, 2020) included in 60,000 plastic formulations available on the market (Shashoua, 2008). The different resin polymers with higher demand worldwide are polypropylene (PP, 19.4%) polyethylene (PE, low (17.4%), and high (12.4%) density), polyvinyl chloride (PVC, 10%), polyethylene terephthalate (PET, 7.9%), and polystyrene (PS, 6.2%), respectively (PlasticsEurope, 2020). Other polymer types account for 26.7% (PlasticsEurope, 2020). Population growth over the last 50 years, rose from 2.5 billion to 7.8 billion people (Worldmeter, 2020), and as plastic production is intrinsically linked, population growth needs to be considered when addressing plastic production.

Evidence of plastics in marine environments was first documented in the late 1960s, early 1970s (Kenyon & Kridler, 1969; Carpenter & Smith, 1972) and while these studies were not targeted at plastic pollution per se, they recognized the potential for widespread contamination effects into the future. Evidence now shows that plastics are ubiquitous in the marine environment, with researchers finding evidence of plastics from the top of Mount Everest to the bottom of the Mariana Trench (Jamieson et al, 2019; Napper et al, 2020). Plastics will prevail in the environment when not properly disposed or recycled, having physical, chemical, and socioeconomic effects and impacts.

Recent studies suggest that we are currently living in the Anthropocene, a geological epoch marked by significant human impact on the Earth's geology, ecosystems, and potentially human health (Slaughter, 2012; Hirschfeld, 2020). Some authors refer to this period as the Plastic Age (Thompson et al, 2009; Napper and Thompson, 2020) and a new lexicon on plastic-related issues has even been recently developed to tackle this global environmental problem (Haram et al, 2020). Plastisphere is one of the terms in the lexicon that describes an ecosystem's ability to adapt and take advantage of environmental plastics, for example as a substrate (Zettler et al, 2013; Amaral-Zettler et al, 2020), and shows how the world is adapting to the presence of plastics in the environment. The term '*microplastics*' under an environmental context, was used for the first time in 2004 to describe the accumulation of microscopic pieces of plastic in the marine environment (Thompson et al, 2004). Several attempts to create an all-inclusive definition for microplastics that include the physicochemical properties of the material have been made in recent years (Frias & Nash, 2019; Hartmann et al, 2019; Rochman et al, 2019) but a consensus has not been reached yet. In this chapter, we will use the definition proposed by Frias and Nash, which is:

> "*Microplastics are any synthetic solid particle or polymeric matrix, with regular or irregular shape and with size ranging from 1 μm to 5 mm, of either primary or secondary manufacturing origin, which are insoluble in water.*"

> (Frias & Nash, 2019)

The three main sources of microplastics in the environment are fibres from domestic washing machines, unaccounted pellets from known leakages, and the distribution of city dust and tyre wear particles (TWP) (Kole et al, 2017; Coyle et al, 2020). Proportionally fibres are more commonly found in environmental samples than both pellets and TWP, yet recent studies show that, a significant contribution to the microplastic concentrations in the environment results from city dust and tyre wear particles (Knight et al, 2020). This chapter focus on the diverse set of methodologies employed to sample, quantify, and identify these microplastics, while acknowledging the global ongoing harmonization and standardization processes.

2.2 TACKLING THE DIVERSITY OF METHODOLOGIES

Since the term microplastics was coined for the first time in 2004, its inclusion in scientific peer-reviewed journals has exponentially grown, reaching a total of 3,200 published to date, with 1,180 published just in 2020 alone. A timeline in relation to plastic publication and the introduction of legislation, CO_2 emissions, and plastic production is provided in Figure 2.1.

The variety of approaches to tackling the issue of microplastics have led to an increase in the knowledge in this area, with the biotic and abiotic related variables being shown to influence the transport, dispersion, and/or accumulation of microplastics in the environment (Franeker & Law, 2015; Kooi et al, 2017; Lebreton et al, 2018; Botterell et al, 2019; Allen et al, 2019). In addition, the extent of this dispersal has led to a wide range of methodologies being deployed for sampling and processing the different environmental matrices (Hidalgo-Ruz et al, 2012; Shim et al, 2017; Lusher et al, 2017; Campanale et al, 2020). The reporting units are another challenge that makes the comparison between studies or geographical areas difficult, if even possible (Frias et al, 2018; Bessa et al, 2019; ISO, 2020). In parallel, the scientific community is striving to reach consensus

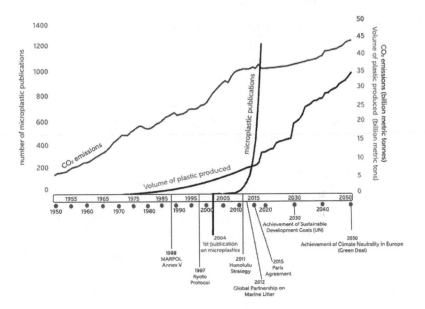

FIGURE 2.1 Timeline of microplastic research and global environmental policy decision. Microplastic publications between 2004 and 2020 (reported on the Web of Science database).

on a single definition for microplastics, while debating what sampling, processing, and data analysis methodologies are 'best practice' to produce high-quality data. However, there is a clear global interest in the quantitative and qualitative methodologies to assess environmental microplastic pollution. This has led to the first step towards international standardization in 2020 with the release of The International Organization for Standardization (ISO) standard for plastics, which also includes microplastics (ISO/TR 21960:2020) (ISO, 2020). It is expected that more standards at national and/or international level will emerge over the next couple of years due to the wide interest in this topic at different stakeholder levels. Given the wide range of environmental matrices (e.g., benthic sediment, terrestrial soil, surface waters in rivers and oceans, wastewater, air, and atmospheric deposition, etc.), and methods to assess them, standardizing the overall microplastic pollution assessment, including reporting units, is often seen as a herculean task.

2.3 LABORATORY PROCESSING

Even within the first methodology review, conducted in 2012 by Valeria Hidalgo-Ruz, it was clear that researchers were already applying different methods to identify and quantify microplastics in the marine environment (Hidalgo-Ruz et al, 2012). Hidalgo-Ruz et al. (2012) highlighted the main techniques and methodologies applied at the time, which largely focused on sampling and laboratory processing, and were divided into four main categories: (1) density separation; (2) filtration; (3) sieving; and (4) visual sorting and identification. The categories highlighted by Hidalgo-Ruz in 2012 are covered, however, the authors provide more emphasis on digestion rather than filtration as this technique has little variation across studies and is largely carried out using a vacuum pump.

2.3.1 Density Separation

Density separation is a common method to extract microplastics from inorganic environmental matrices. This method is based on the fact that plastics have low density, and it follows the principle that plastics will float to the surface of a high-density solution, where the supernatant can be easily filtered to retrieve plastic particles. There are many examples described in the literature that show the wide range of polymer densities that are in use today (Frias et al, 2018; Bessa et al, 2019). Hidalgo-Ruz et al, 2012 reported in their review that plastic density ranges between 0.8 to 1.4 g cm^{-3} and that the density solutions used at the time ranged from 1.2 to 1.4 g cm^{-3} to recover plastic particles. Recent reports highlight that there are polymers whose densities are greater than 1.4 g cm^{-3} (e.g., polytetrafluoroethylene has a density that ranges from 2.1 to 2.3 g cm^{-3}) (Frias et al, 2018), which were not commonly found in earlier studies, potentially due to the density solutions used at the time. Therefore, alternative solutions with higher densities were identified to allow extraction of microplastic particles from the different environmental matrices (Frias et al, 2018; Imhof et al, 2012). Some of these density solutions, such as lithium metatungstate $(Li_6(H_2W_{12}O_{40}))$, zinc chloride $(ZnCl_2)$, zinc bromide $(ZnBr_2)$, or sodium iodide (NaI), which have densities of 1.6, 1.8, 1.71, 1.80 g cm^{-3}, respectively, are able to recover a higher proportion of microplastics (Frias et al, 2018), and provide a more comprehensive assessment of the sample in question. Due to the ongoing harmonization and standardization processes, a consensus on which density separation method is the most appropriate for each environmental matrix has not been reached yet. There are a considerable number of factors to consider, from the toxicity of the solution to economic factors. However, identifying the most appropriate techniques within the methodologies, i.e., 'best practice', for each environmental matrix, is still a challenging task.

2.3.2 Digestion

After sampling of an environmental matrix (e.g. water, air, sediment, or biota), a range of processing methods are being applied (Frias et al, 2018; Filgueiras et al, 2019; Bessa et al, 2019). Depending on the nature of the matrix there are many factors influencing the selection of methods to be used during processing (Enders et al, 2020). Sediments usually undergo a density separation process, while biota follows an alkaline digestion process (usually using a 10% potassium hydroxide (KOH) solution); however, a mix of density separation and alkaline digestion has been recently proposed (Enders et al, 2020). For water samples, depending on the nature of the study and on how pure the water sample is, both methods can be applied. For example, if a water sample contains high organic matter content, the processing process usually involves alkaline digestion. A recent study from Enders et al. (2020), suggested that both density separation and alkaline digestion can be conducted to ensure that all microplastic particles are extracted successfully from the samples. Other authors suggest conducting an enzymatic purification of environmental samples to retrieve microplastics (Löder et al, 2017), but these processes tend to be more expensive and thus not widely applied. Initially, studies did not report how the supernatant was extracted from the matrices with the focus solely on the pore size of the fibreglass filter papers, which ranged from 1 to 2 µm (Hidalgo-Ruz et al, 2012). Today, filtration methods are consistently specified within the methodologies section of published manuscripts. In addition, the wide range of filters applied, including stainless-steel filters, are also presented with the selected microplastic identification techniques in mind (Primpke et al, 2020). For biota samples, the digestion process, i.e., chemical for digestion, needs to consider both the size of the organism and its lipid content (Lusher et al, 2017). Microplastic studies focussed on ingestion largely target the digestive tract as the point of intake (Hara et al, 2020), although there are a number of studies where the entire organism is digested to assess both ingestion and uptake by other routes, e.g., respiratory system (Woods et al, 2018; Roch et al, 2020).

2.3.3 Sieving

Sieving is usually performed in sediment and surface water samples, to separate the larger natural elements (e.g. wood, algae, etc.) from macro-, meso- and microplastics, while consequently dividing them into different size fractions. This process, although relatively simple, is important to (a) understand the microplastic size scale in a given region; (b) allow comparison between studies, and (c) improve mathematical models. Understanding fragmentation and degradation processes in the environment allow us to assess plastics and microplastics in several size classes, potentially creating mechanisms to collect them. Assessing fragmentation and degradation processes also allow the models to predict microplastic concentration inputs to ecosystems (Kooi et al, 2016; Besseling et al, 2019). Size fractions are separated using stainless-steel mesh sieves, which usually range from 0.038 to 4.75 mm, with 1 mm the most common size mesh reported (Hidalgo-Ruz et al, 2012). Sieving and pre-sieving of environmental samples are routinely applied where there is a high biomass content, as not removing excess biomass will hinder sorting and identification processes down the line. A new sieving method targeted at minimizing microplastic loss while transferring samples from the sampling to the processing vials has been recently published (Nakajima et al, 2019). This method uses a stainless-steel sieve that is adapted to sit on top of glass beakers and has a 32 µm size mesh that allows the retrieval of plastic particles within 2 size classes (100–500 µm and 500–1,000 µm). The method has a higher efficiency (~99%) when compared to the filtration method (88–97%) and it contributes to reducing the processing time of the subsequent digestion (Nakajima et al, 2019).

2.3.4 Visual sorting and identification

The last category identified in the review from Hidalgo-Ruz was visual sorting and identification of microplastic particles, which is the field that has had the most technological advancements since 2012. Initial studies on microplastics were mainly focussed on the larger microplastics (>1 mm in diameter), which were relatively easy to find and identify in an environmental matrix (Hidalgo-Ruz et al, 2012). Visual identification was and is still a common method for larger items, but other identification methods have since been developed or implemented to aid the identification of the smaller microplastics. Nowadays, there is a non-formal agreement that the microplastic upper limit does not extend beyond 5 mm with researchers now referring to larger plastics items as meso- or macroplastics (van Cauwenberghe et al, 2015). Identification techniques for microplastics have gone beyond the realm of sole visual identification (Fischer and Scholz-Böttcher, 2017; Primpke et al, 2020; Enders et al, 2020) with several identification techniques available, such as RAMAN

Spectroscopy, Fourier Transform Infrared Spectroscopy (FTIR) or Pyrolysis-Gas Chromatography-Mass Spectrometry (Py-GC-MS). All these automatic techniques reduce human error from visual identification and allow the smaller size range to be included in the assessment thus, providing a better overall appreciation of microplastic concentrations in the environment.

Spectroscopic techniques such as RAMAN and FTIR or spectrometric techniques such as Py-GC-MS, are commonly used to determine the chemical spectra or the mass of polymers in environmental samples, respectively. These methods allowed the field of microplastic research to identify smaller sizes (10–20 μm), showing that the initial scale of microplastic pollution was largely underestimated. In relation to the spectroscopic methods, the debate in the scientific community has evolved to acknowledge that both FTIR and RAMAN spectroscopy are similar techniques in that they provide the same outcome (Cabernard et al, 2018; Xu et al, 2019). The inclusion of mass spectrometry as a complementary method to spectroscopy is also gaining more followers, as it has the capacity to report the mass of microplastics present, therefore providing a more comprehensive summary of the microplastic pollution within any one environmental matrix (Fischer and Scholz-Böttcher, 2017).

Progress in the field of microplastic identification is evolving fast (Primpke et al, 2020) with identification software being constantly developed and/or updated, and with methods constantly being developed to improve quality assurance and quality control (QA/QC) related to laboratory processing of microplastics. Furthermore, as sources and pathways of microplastic pollution in the environment are still being identified, studies will begin to focus on these relatively unknown aspects, e.g., fibres, pellets, or TWP (Napper & Thompson, 2016; Karlsson et al, 2018; Knight et al, 2020). The main challenge with TWP is its quantification, due to the apparent invisibility of such particles under spectroscopic techniques, which is due to the colour and composition of TWP.

2.4 METHODOLOGIES

With the rapid methodological advancements in this field and a variety of environmental matrices to be assessed, researchers had to adapt known processes and techniques while devising multiple-layer methods to ensure optimal recovery of microplastics (Enders et al, 2020). Most methods are targeted at the key environmental matrices (sediment, water, and biota), while a few studies are starting to focus on a relatively unexplored matrix, air (Zhang et al, 2020). Assessing microplastic concentrations requires a variety of processing techniques for the particular organic or inorganic matrix being explored (Enders et al, 2020). Proposed methodologies targeted at surface water (Filgueiras et al, 2019), intertidal and benthic sediments (Frias et al, 2018), and biota (Bessa et al, 2019) have previously been published as part of international harmonization and inter-calibration exercises in European projects such as BASEMAN or EPHEMARE (Frias et al, 2018; Filgueiras et al, 2019; Bessa et al, 2019). In Europe, other inter-calibration exercises organized by the European Commission Joint Research Centre are currently ongoing (JRC, 2020). Such protocols and exercises stress the need for reliable and comparable data. The protocols from BASEMAN and EPHEMARE provide a comprehensive overview of sampling, processing, and the identification of microplastics in the three main environmental matrices. In addition, they include a list of materials and easy step-by-step methodology (Frias et al, 2018; Filgueiras et al, 2019; Bessa et al, 2019). Although there are as yet no standardized procedures for each individual matrix, it is important to understand that methodologies based on these 'best practice' protocols can be, and have been adopted to case studies worldwide including Ireland with targeted studies on the surface waters (Frias et al, 2020), on benthic sediment (Pagter et al, 2020), and on biota (Hara et al, 2020). Each matrix necessitates the researcher to follow specific processes, for example, surface seawater or biota often require alkaline digestion using 10% KOH (Frias et al, 2020; Hara et al, 2020), while benthic sediment required a sodium tungstate dihydrate solution to be used as a density separation technique (Pagter et al, 2020). The initial assessment of microplastic concentrations is important to provide baselines for the abundance of microplastics for given geographical areas or species. Monitoring of those ecosystems and species will lead to a better understanding of the inputs and accumulation areas, where long time-series which in turn can provide a comprehensive overview on the trends at a population level. Baselines for commercial species are important to assess the potential human exposure to microplastics through ingestion of seaweed, fish, and shellfish. For example, a targeted microplastic study on the Dublin Bay Prawn (*Nephrops norvegicus*) was conducted in Ireland, to estimate the potential ingestion of microplastics by human consumers (Hara et al. 2020). It is important to note that to the author's knowledge, no human health impacts have been reported within the scientific literature, and most reviews only refer to potential effects (SAPEA, Science Advice for Policy by European Academies, 2019; Campanale et al, 2020; GESAMP, 2020).

2.5 METRICS

Another important aspect to take into consideration is the comparison of metrics. Conducting literature reviews focussed on how comparing results is challenging, if even a possible task, due to the variety of methods applied and to the wide range of reporting units in the literature (Hidalgo-Ruz et al, 2012). Together with the lack of reporting standards, it is common for reporting units to be expressed in the most convenient way to a specific environmental matrix or to the background of multidisciplinary teams. One of the earlier studies that compared the wide range of methodologies to assess microplastic contamination, highlighted the impossibility of comparing different studies which used different sampling, processing, and data analysis methods (Hidalgo-Ruz et al, 2012). In that study, there were already efforts to standardize microplastic concentrations by area and/or volume. It is common to have a wide range of report units depending on the environmental matrix, some common examples include: the concentrations of microplastics (MP) expressed by area (#MP m^{-2} or by #MP in 100 m linear), by volume (#MP m^{-2}, #MP L^{-1}), by mass (either in kg, or expressed by a standard amount of dry sediment and usually reported as #MP/100g dry weight) or, in relation to airborne particles, results are expressed by fallout abundance (#MP/ m^{-2} d^{-1}) (Frias et al, 2018; Bessa et al, 2019, Zhang et al, 2020). Reporting microplastic concentrations per area or volume is not seen as a challenge as data conversions are easy; however, conversions might be misleading if authors do not explain their calculations within their manuscripts.

Due to the advent of spectrometric techniques, it is also common to report the mass of microplastics per volume or per area (i.e., g MP m^{-2} or g MP m^{-3}), which is a positive addition but unlikely to be as widely published due to a limited number of the Pyr-GC-MS machines being used to assess microplastics. Efforts have also been made to compare diverse methodologies to sample the same environmental matrix under similar conditions in the field (Pagter, Frias, and Nash, 2018) or to test the efficacy of novel tools to assess microplastic concentrations (Zobrok and Esiukova, 2017). These comparisons, along with inter-calibration exercises coordinated by research groups will contribute to the identification of the best methodologies.

2.6 CURRENT AND FUTURE CHALLENGES TO BE ADDRESSED

Plastic is a versatile material and the characteristics that make it suitable for a wide range of uses (e.g., lightweight, resistance to corrosion, low thermal conduction, low electrical conduction, and low cost) are the same that result in plastics being harmful to the environment, to aquatic organisms and to local economies

(Bockhorn et al, 1999, Rochman et al, 2013, Wright et al, 2013, UNEP, 2016). However, a distinction should be made between essential and non-essential single-use plastic materials, and this aspect has been highlighted within the current global coronavirus outbreak. For example, the use of disposable personal protective equipment (PPE) within the healthcare system is not only a necessity but an important barrier in curbing the spread of the virus, while the public has the potential to avail of reusable masks many are choosing single-use masks. There are essential single-use items that are needed in the research and medicine fields because there are no alternatives available. By contrast, non-essential single-use items (e.g., straws, plastic cutlery, etc), need to be reconsidered through an eco-design perspective that includes circular economy.

Methodologies into the future will continue to apply a wide range of divergent techniques to assess microplastics in environmental matrices due to the lack of standardization until a consensus is agreed. It is important to understand that there is no single solution to tackle the sampling, processing, and identification of microplastics in these environmental matrices as they depend on a variety of factors.

Research in this field continues to flourish and expand to understand sources, pathways, and the processes that influence fragmentation, biofouling, leaching of chemicals, and potential health impacts to marine organisms, humans, and ecosystems (Brennecke et al, 2016; Fazey & Ryan, 2016). Technical, conceptual, and logistic limitations, as well as the lack of standardized procedures, are the main limiting factors in tackling knowledge gaps in this field. The scientific community is constantly reinventing itself to address different aspects associated with methodologies, and it is expected that new modeling techniques and machine-learning systems targeted at identifying microplastics in the environmental samples will become common methodologies in future studies as soon as the next decade.

REFERENCES

Amaral-Zettler, L., Zettler, E. & Mincer, T. (2020) Ecology of the plastisphere. *Nature Reviews Microbiology*, 18, 139–151. 10.1038/s41579-019-0308-0

Allen, S., Allen, D., Phoenix, V., La Roux, G., Jimenez, P. D., Simonneau, A., Binet, S. & Galop, D. (2019) Atmospheric transport and deposition of microplastics in a remote mountain catchment. *Nature Geoscience*, 12, 339–344. 10.1038/s41561-019-0335-5

Botterell, Z., Beaumont, N., Dorrington, T., Steinke, M., Thompson, R. & Lindeque, P. (2019) Bioavailability and effects of microplastics on marine zooplankton: A review. *Environmental Pollution*, 245, 98–110. 10.1016/j.envpol.2018.10.065

Bockhorn, H., Hornung, A., Hornung, U. & Schawaller, D. (1999) Kinetic study on the thermal degradation of polypropylene and polyethylene. *Journal of Analytical and Applied Pyrolysis* 48, 93–109. 10.1016/S0165-2370(98)00131-4

Bessa et al. (2019) Harmonized protocol for monitoring microplastics in biota. *JPI Oceans*. 10.13140/RG.2.2.28588.72321/1

Besseling, E., Redondo-Hasselerharm, P., Foekema, E. & Koelmans, A. (2019) *Quantifying ecological risks of aquatic micro- and nanoplastics. Critical reviews in Environmental Science and Technology.* Francis and Taylor Online. 10.1080/10643389.2018.1531688

Brennecke, D., Duarte, B., Paiva, F., Caçador, I. & Canning-Clode, J. (2016) Microplastics as a vector for heavy metal contamination from the marine environment. *Estuarine, Coastal and Shelf Science*, 178, 189–195. 10.1016/j.ecss.2015.12.003

Cabernard, L., Roscher, L., Lorenz, C., Gerdts, G. & Primpke, S. (2018) Comparison of Raman and Fourier transform infrared spectroscopy for the quantification of microplastics in the aquatic environment. *Environmental Science & Technology*, 52, 22, 13279–13288. 10.1021/acs.est.8b03438

Campanale, C., Massrelli, C., Savino, I., Locaputo, V. & Uricchio, V. F. (2020) A detailed review study on potential effects of microplastics and additives of concern on human health. *International Journal of Environmental Research and Public Health*, 17(4), 1212 10.3390/ijerph17041212

Campanale, C., Savino, I., Pojar, I., Massarelli, C. & Uricchio, V. F. (2020) A practical overview of methodologies for sampling and analysis of microplastics riverine environments. *Sustainability 2020*, 12, 6755; 10.3390/su12176755

Carpenter, E. & Smith, K. (1972) Plastics on the Sargasso Sea Surface, *Science*, 175(4027), 1240–1241. 10.1126/science.175.4027.1240

Coyle, R., Hardiman, G. & O'Driscoll, K. (2020) Microplastics in the marine environment: A review of their sources, distribution processes and uptake into ecosystems. *Case Studies in Chemical and Environmental Engineering*, 10010. 10.1016/j.cscee.2020.100010

Enders, K., Lenz, R., Ivar do Sul, J., Tagg, A. & Labrenz, M. (2020) When every particle matters: A QuEChERS approach to extract microplastics from environmental samples. *MethodsX*, 7, 100784. 10.1016/j.mex.2020.100784

Fazey, F. M. C. & Ryan, P. G. (2016) Biofouling on buoyant marine plastics: An experimental study into the effect of size on surface longevity. *Environmental Pollution* 210, 354–360. 10.1016/j.envpol.2016.01.026

Franeker, J. A. & Law, K. L. (2015) Seabirds, gyres and global trends in plastic pollution. *Environmental Pollution*, 203, 89–96. 10.1016/j.envpol.2015.02.034

Filgueiras et al. (2019) Standardised protocol for monitoring microplastics in seawater. *JPI Oceans*. 10.13140/RG.2.2.14181.45282

Fischer, M. & Scholz-Böttcher, B. (2017) Simultaneous trace identification and quantification of common types of microplastics in environmental samples by pyrolysis-gas chromatography–mass spectrometry. *Environmental Science & Technology*, 51 (9), 5052–5060. 10.1021/acs.est.6b06362

Frias et al. (2018) Standardised protocol for monitoring microplastics in sediments. *JPI Oceans*. 10.13140/RG.2.2.36256.89601/1

Frias, J. & Nash, R. (2019) Microplastics: Finding a consensus on the definition. *Marine Pollution Bulletin*. 138, 145–147. 10.1016/j.marpolbul.2018.11

Frias, J., Lyashevska, O., Joyce, H., Pagter, E. & Nash, R. (2020) Floating microplastics in a coastal embayment: A multifaceted issue. *Marine Pollution Bulletin*, 158, 111361. 10.1016/j.marpolbul.2020.111361

GESAMP (2020) Proceedings of the GESAMP International Workshop on assessing the risks associated with plastics and microplastics in the marine environment (Kershaw, P.J., Carney Almroth, B., Villarrubia-Gómez, P., Koelmans, A.A., and Gouin, T., eds.). (IMO/FAO/UNESCO-IOC/UNIDO/WMO/IAEA/UN/ UNEP/ UNDP/ISA Joint Group of Experts on the Scientific Aspects of Marine Environmental Protection). Reports to GESAMPNo. 103, 68 pp. http://www.gesamp.org/publications/gesamp-international-workshop-on-assessing-the-risks-associated-with-plastics-and-microplastics-in-the-marine-environment

Hartmann, N., Hüffer, T. & Thompson, R.C., et al. (2019) Are we speaking the same language? Recommendations for a definition and categorization framework for plastic debris. *Environ. Sci. Technol.*, 2019, 53(3), 1039–1047. 10.1021/acs.est.8b05297

Haram, L., Carlton, J., Ruiz, G. & Maximenko, N. (2020) Marine pollution bulletin, 150, 110714. 10.1016/j.marpolbul.2019.110714

Hara, J., Frias, J. & Nash, R. (2020) Quantification of microplastic ingestion by the decapod crustacean *Nephrops norvegicus* from Irish waters. *Marine Pollution Bulletin*, 152, 110905. 10.1016/j.marpolbul.2020.110905

Harari, Y. N. (2015) *Sapiens: A Brief History of Humankind.* Harper. ISBN: 978-0062316097

Hidalgo-Ruz, V., Gutow, L., Thompson, R. C. & Thiel, M. (2012) Microplastics in the marine environment: a review of the methods used for identification and quantification. *Environmental Science and Technology*, 46 (2012), 3060–3075. 10.1021/es2031505

Hirschfeld, K. (2020) Microbial insurgency: Theorizing global health in the Anthropocene. *The Anthropocene Review*, 7(1), 3–18, 10.1177/2053019619882781

Imhof, H. K., Schmid, J., Niessner, R., Ivleva, N. P. & Laforsch, C. (2012), A novel, highly efficient method for the separation and quantification of plastic particles in sediments of aquatic environments. *Limnology and Oceanography Methods*, 10, 524–537. 10.4319/lom.2012.10.524

ISO (2020) Plastics - environmental aspects - state of knowledge and methodologies. ISO/TR 21960:2020. https://www.iso.org/standard/72300.html

Jamieson, A., Brooks., L., Reid, W., Piertney, S., Narayanaswamy, B. & Linley, T. (2019) Microplastics and synthetic particles ingested by deep-sea amphipods in six of the deepest marine ecosystems on Earth. *R. Soc. open sci.* 6, 180667. 10.1098/rsos.180667

JRC, 2020. European Commission website. (Consulted in December 2020) https://ec.europa.eu/jrc/en/news/finding-right-methods-measuring-microplastics-water

Karlsson, T., Arneborg, L., Broström, G., Almroth, B., Gipperth, L. & Hassellöv, M. (2018) The unaccountability case of plastic pellet pollution. *Marine Pollution Bulletin*, 129 (1), 52–60. 10.1016/j.marpolbul.2018.01.041

Knight, L., Parker-Jurd, F., Al-Sid-Cheikh, M. & Thompson, R. (2020) Tyre wear particles: An abundant yet widely unreported microplastics? *Environmental Science and Pollution Research*, 27, 18345–18354 10.1007/s11356-020-08187-4

Kenyon, K. W. & Kridler, E. (1969) Laysan albatrosses swallow indigestible matter, *The Auk*, 86, 2. 10.2307/4083505

Kole, P. J., Löhr, A. J., Van Belleghem, F. G. A. J. & Ragas, A. M. J. (2017) Wear and Tear of Tyres: A Stealthy Source of Microplastics in the Environment. *International Journal of Environ Research and Public Health*, 14(10), 1265. 10.3390/ijerph14101265

Kooi, M., Nes, E., Scheffer, M. & Koelmans, A. (2017) Ups and downs in the ocean: Effects of biofouling on vertical transport of microplastics. *Environmental Science and Technology*, 51 (14), 7963–7971. 10.1021/acs.est.6b04702

Kooi, M., Reisser, J., Slat, B., Ferrari, F. F., Schmid, M. S., Cunsolo, S., &…., Koelmans, A. A. (2016) The effect of particle properties on the depth profile of buoyant plastics in the ocean. *Scientific Reports*, 6(1), 33882. 10.1038/srep33882

Lebreton, L., Slat, B., Ferrari, F. et al. (2018) Evidence that the great pacific garbage patch is rapidly accumulating plastic. *Sci. Rep.*, 8, 4666 10.1038/s41598-018-22939-w

Löder, M., Imhof, H., Ladehoff, M., Löschel, L., Lorenz, C., et al. (2017) Enzymatic purification of microplastics in environmental samples. *Environ. Sci. Technol.*, 2017, 51(24), 14283–142092 10.1021/acs.est.7b03055

Lusher, A., Welden, N., Sobral, P. & Cole, M. (2017) Sampling, isolating and identifying microplastics ingested by fish and invertebrates. *Analytical Methods*, 9, 1346–1360. 10.1039/C6AY02415G

Martinho, M.G. e & Cunha, F. (2010) Practical Guide for Waste Management. Verlag Dashofer (ISBN 978-972-8906-07-8). https://novaresearch.unl.pt/en/publications/manual-pr%C3%A1 tico-para-a-gest%C3%A3o-de-res%C3%ADduos

Napper, I. & Thompson, R. (2016) Release of synthetic microplastic plastic fibres from domestic washing machines: Effects of fabric type and washing conditions. *Marine Pollution Bulletin*, 112 (1-2), 39–45. 10.1016/j.marpolbul.2016.09.025

Napper, I., Davies, B., Clifford, H. & Elvin, S. et al. (2020) Reaching new heights in plastic pollution - preliminary findings of microplastics on Mount Everest. *CellPress*, 3 (5), 621–630. 10.1016/j.oneear.2020.10.020

Nakajima, R., Lindsay, D., Tsuchiya, M., Matsui, R., Kitahashi, T., Fujikura, K. & Fukushima, T. (2019) A small, stainless-steel sieve optimized for laboratory breaker-based extraction of microplastics from environmental samples. *MethodsX*, 6, 1677–1682. 10.1016/j.mex.2019.07.012

Pagter, E., Frias, J. & Nash, R. (2018) Microplastics in Galway Bay: A comparison of sampling and separation methods. *Marine Pollution Bulletin*, 135, 932–940. 10.1016/j.marpolbul.2018.08.013

Pagter, E., Frias, J, Kavanagh, F. & Nash, R. (2020) Varying levels of microplastics in benthic sediments within a shallow coastal embayment. *Estuarine, Coastal and Shelf Science*, 243, 106915. 10.101 6/j.ecss.2020.106915

PlasticsEurope (2010) Plastics – the Facts, 2010: An analysis of European plastics production, demand, and recovery for 2009. https://www.plasticseurope.org/en/resources/publications/171-plastics-facts-2010

PlasticsEurope (2020) Plastics – the Facts, 2020: An analysis of European plastics production, demand, and waste data. https://www.plasticseur ope.org/en/resources/publications/4312-plastics-facts-2020

Primpke, S., Cross, R., Mintenig, S., et al. (2020) Toward the systematic identification of microplastics in the environment: evaluation of a new independent software tool (siMPle) for spectroscopic analysis. *Applied Spectroscopy*, 74(9), 1127–1138. 10.1177/0003702820917760

Rochman, C., Brookson, C., Bikker, J., Djuric, N., et al. (2019) Rethinking microplastics as a diverse contamination suite. *Environmental Toxicology and Chemistry*, 38(4), 703–711, 2019 10.1002/etc.4371

Roch, S., Friedrich, C. & Brinker, A. (2020) Uptake routes of microplastics in fishes: Practical and theoretical approaches to test existing theories. *Scientific reports*, 10, 3896 (2020). 10.1038/s41598-020-60630-1

Slaughter, R. (2012) Welcome to the Anthropocene. *Futures*, 44(2), 119–126, Elsevier. 10.1016/j.futures.2011.09.004

SAPEA, Science Advice for Policy by European Academies. (2019) *A Scientific Perspective on Microplastics in Nature and Society*. Berlin: SAPEA. 10.26356/microplastics

Shashoua, Y. (2008) *Conservation of Plastics, - materials science, degradation, and preservation*. Amsterdam: Elsevier/Butterworth-Heinemann. ISBN: 978-0-7506-6495-0.

Shim, W. J., Hong, S. H. & Eo, S. E. (2017) Identification methods in microplastic analysis: A review. *Analytical Reviews*, 9, 1384–1391. 10.1039/C6AY02558G

Thompson, R. C., Olsen, Y., Mitchell, R. P., Davis, A., Rowland, S. J., John, A. W., McGonigle, D. & Russell, A. E. (2004). Lost at sea: where is all the plastic? Science, 304(5672), 838. 10.1126/science.1 094559.

Thompson, R., Swan, S., Moore, C. & vom Saal, F. (2009) Our plastic age. Philosophical transactions of the royal society B, *Biological Sciences*, 364, 1526. 10.1098/rstb.2009.0054

UNEP (2016) *Marine plastic debris and microplastics – Global lessons and research to inspire action and guide policy change.* United Nations Environment Programme, Nairobi. ISBN: 978-92-807-3580-6. http://hdl.handle.net/20.500.11822/7720

van Cauwenberghe, L., Devriese, L., Galgani, F., Robbens, J. & Janssen, C. (2015) Microplastics in sediments: A review of techniques, occurrences and effects. *Marine Environmental Research*, 111, 5–17. 10.1016/j.marenvres.2015.06.007

Woods, M., Stack, M., Fields, D., Shaw, S. & Matrai, P. (2018) Microplastic fiber uptake, ingestion and egestion rates in the blue mussel (*Mytilus edulis*). *Marine Pollution Bulletin*, 137, 638–645. 10.1016/j.marpolbul.2018.10.061

Worldmeter (2020) World population. Website consulted on 15/12/ 2020: https://www.worldometers.info/world-population/

Wright, S. L., Thompson, R. C. & Galloway, T. S. (2013). The physical impacts of microplastics on marine organisms: A review. *Environmental Pollution*, 178, 483–492. 10.1016/j.envpol.2013.02.031.

Xu, J., Thomas, K., Luo, Z. & Gowen, A. (2019) FTIR and Raman imaging for microplastics analysis: State of the art, challenges and prospects. *TrAC Trends in Analytical Chemistry*, 119, 115629. 10.1016/j.trac.2019.115629

Zettler, E., Mincer, T. & Amaral-Zettler, L. (2013) Life in the "Plastisphere": Microbial communities on plastic marine debris. *Environmental Science & Technology*, 47(13), 7137–7146, 10.1021/es401288x

Zobrok, M. & Esiukova, E. (2017) Evaluation of the Munich Plastic Sediment Separator efficiency in extraction of microplastics from natural marine bottom sediments. *Limnology and Oceanography: Methods* 15(2017), 967–978. 10.1002/lom3.10217

Zhang, Y., Kang, S., Allen, S., Allen, D., Gao, T. & Silanpää, M. (2020) Atmospheric microplastics: A review on current status and per-spectives. *Earth-Science Reviews*, 203, 103118. 10.1016/j.earscirev.202 0.103118

Understanding the Source, Behaviour, and Fate of Nanoplastics in Aquatic Environments

Timothy Sullivan[1,2] and Irene O'Callaghan[1,3]

[1]School of Biological, Earth and Environmental Sciences, University College Cork, Ireland
[2]Environmental Research Institute, 6 Lee Rd, University College Cork, Ireland
[3]School of Chemical Sciences, University College Cork, Ireland

CONTENTS

DOI: 10.1201/9781003109730-3

3.1 INTRODUCTION

Incineration aside, it has been proposed that most plastics ever produced are still present in the environment in some form (Barnes et al, 2009). It should then be no surprise that plastic pollution is now well-established as being almost ubiquitous (or omnipresent) in aquatic environments at all spatial scales (Boyle & Örmeci, 2020; Strungaru et al, 2019). While nano and microplastics research has received most effort to date in marine environments, there is also an increasing focus on understanding exposure, uptake by organisms, the impact, and ultimate fate of plastic debris in freshwater environments (Gigault et al, 2021; Horton et al, 2017). Some trends have started to emerge in relation to dependencies and differences in particulate size in marine environments for example. Nevertheless, the precise impact and broad implications of plastic pollution in aquatic ecosystems under wide-ranging environmental conditions are still very much under investigation (Wang et al, 2021).

Investigation of plastic pollution in aquatic environments now increasingly focuses on the hypothetical prevalence and potential impacts of nano-sized plastic fragments (Koelmans et al, 2015). When addressing this research, it should be emphasized again that the terminology for determining plastic size categorization has been poorly defined in the literature to date. This is primarily as a result of the common definition of 'microplastic', which is often taken to apply to particles of up to 5 mm in length along the shortest dimension, but also relating to varying chemical compositions, associations, and polymer densities for example. Consequently, the upper range of 'nanoplastics' has been set at anywhere between 100 nm and 335 µm, depending on the source (Hartmann et al, 2019; Mitrano et al, 2021). Among these definitions is the attempt to align nanoplastic definitions with that of legislation surrounding engineered nanomaterials (<100 nm) and the opinion that the definition should reflect the ecological definition of nanoplankton (<20 um) (Wagner & Reemtsma, 2019).

This chapter focuses on nanoplastics defined as nano-sized particles of plastic composition with at least one dimension less than 1 μm, as this definition has met with approval from numerous authors (da Costa, 2018a). Many points discussed here will, however, also find relevance in the context of low μm-scale plastics, as the properties of these materials evolve gradually as dimensions reduce towards nano-scale (Donaldson & Poland, 2013).

Although detection and quantification in any environment have proven challenging to date (Shen et al, 2019), nanoplastics are likely present in the environment as a result of both direct release and degradation of larger plastics through a variety of biological and nonbiological processes (Huang et al, 2021; Mateos-Cárdenas et al, 2020; Mattsson et al, 2018), and can be affected by a wide range of transport mechanisms (Huang et al, 2021). As has been observed with microplastics, the underlying mechanisms of any toxicity of nanoplastics can be complex, and detailed understanding of the latter suffers from the relative novelty of the field (Wagner & Reemtsma, 2019), but is undergoing rapid development (Kukkola et al, 2021). The recent flurry of activity has produced a general picture of the possible behaviour of nanoplastics in aquatic environments (Figure 3.1), and it seems likely that their physical behaviour may not be entirely dissimilar to other nanomaterials in some regards.

3.2 PROCESSES AND MECHANISMS

Nanoplastics deserve consideration as distinct from, but combining the intrinsic characteristics of, both meso- and micro-scale plastics and non-plastic nanomaterials (Gigault et al, 2021; Mitrano et al, 2021). Nanoplastics, due to their composition, are regarded as biochemically inert, but their nano-specific surface chemistry allows for complexation and vectoring of other contaminants and pathogens, and engineered particles often contain additives that may enter the environment (da Costa, 2018a;

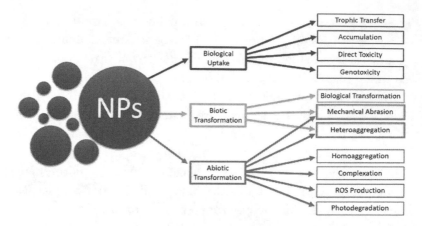

FIGURE 3.1 A summary of the various recognized mechanisms of transformation that nanoplastics have been seen to undergo in aquatic environments.

Markiewicz et al, 2018; Wang et al, 2021). Whether nanoplastics can act as significant environmentally relevant pathways for sorbed pollutants is still unclear (Kukkola et al, 2021). Similarly, their size increases the likelihood that these particles will be ingested and accumulated, as well as potentially allowing them to traverse the blood-brain barrier (Almutairi et al, 2016; da Costa, 2018a; Kashiwada, 2006). Furthermore, it has been noted that nanoplastics may be classed as 'primary' or 'secondary,' wherein the former enters the environment as a nano-scale plastic particle, while the latter is the product of larger plastic degrading within the environment, leading to a nano-scale particle (da Costa, 2018a; González-Pleiter et al, 2019).

3.2.1 The Importance of Scale

As previously mentioned, a formal definition of what constitutes a nanoplastic is still lacking (da Costa, 2018b), with some published works applying the term to μm-scale particles. It is important to appreciate that the chemical reactivity of smaller-scale particles depends on the surface area of the particle, which, for a spherical particle, is proportional to the square of its diameter. An example of this relationship is the reaction of polystyrene nanoplastics (PS-NPs) with the hydroxyl radical (OH˙) (Bianco et al, 2020). As the volume of this particle is proportional to the cube of its diameter, the reactivity per unit mass of this spherical particle is proportional to the inverse of its diameter. Thus, a factor of 10 reduction in the size of the particle corresponds to a factor of 10 increase in the relative reactivity of the particle. It follows that the correct definition of particle size is not simply a matter of categorization, but essential for understanding the potential environmental reactivity of a particle.

To provide a concrete example, Hartmann et al. (2019) cites two publications, each considering spherical plastic particles of diameter 335 μm, but labelling them microplastics and nano-plastics, respectively. These papers are not just being inconsistent in their use of terminology: choosing to define the nanoplastic size limit as 100 nm or 335 μm changes the relative reactivity of a typical "nanoplastic" by a factor of 3,350. Therefore, it is of the greatest importance that terminology is defined and understood and ideally standardized across future studies where possible.

In defining nanoplastics as having a dimension less than 1 μm, this chapter focuses on the more reactive end of the scale, as this is the area of most interest from the point of view of evaluating the reactive nature of nano-scale plastic particles.

3.2.2 Zeta Potential

Zeta Potential (ZP) is often used as a proxy for colloidal stability in aquatic systems, with a region of instability estimated as lying between ±30 mV, so it would follow that those interactions between cells and nanoplastics may be strongly dependent on ZP. A large difference between the ZP of a nanoplastic and that of a cell is taken to promote aggregation of these two bodies, a process commonly termed heteroaggregation (Oriekhova & Stoll, 2018).

The use of ZP for determining the prevalence of hetero-aggregation is supported by the DLVO theory, which posits that the conditions for aggregation depend on the contrasting van der Waals and electrostatic forces, the latter quantified by ZP. Munk et al. (2017) emphasizes that ZP is strongly pH-dependent, which would suggest different heteroaggregation behaviour for different pH conditions (Munk et al, 2017).

The relative surface charges of the nanoplastic and the cell membrane characterize the probability of heteroaggregation, and this may have implications for the resulting toxicity (Oriekhova & Stoll, 2018). This is especially true in the case of nanoplastics, where direct contact is supposed to be the primary mechanism of action (Khoshnamvand et al, 2021). As with other nanoparticles, nanoplastic ZP has been found to correlate with growth inhibition of the freshwater cyanobacteria *Synechococcus elongatus*, with positively charged nanoplastics strongly inhibiting growth of the negatively-charged bacteria, whereas no growth inhibition effect was observed in the case of negatively-charged nanoplastics (Feng et al, 2019). Similarly, nanoplastics with a negative zeta potential were found to exhibit very little toxicity to the algae *Scenedesmus obliquus* under a range of environmental conditions (Liu et al, 2019). Surface charge can have a different effect altogether in the context of plant uptake, as particle uptake by plants has been seen to be charge-dependent with pathways depending on the polarity of the charge (Sun et al, 2020).

These observations indicate that ZP is a useful proxy for measuring the potential for heteroaggregation of a given nano-plastic to the cell membrane. It is worth emphasizing, however, that the process of heteroaggregation is not in itself a mechanism of toxicity. Heteroaggregation may result in either physical da-mage to the cell membrane or direct toxicity resulting from the chemical composition of the nanoplastic or its complexation to another toxic compound. Additionally, it is worth noting that the relatively small body of work on this subject to date has been largely restricted to uni-cellular organisms, so it is not clear how these findings would translate to the context of multi-cellular organisms.

3.2.3 Interactions between Organic Matter and Nanoplastics

As pointed out by Cai et al., complex environmental factors such as variation in pH, natural organic matter, suspended clay particles, or presence of bacteria can widely influence the behaviour of nanoplastics once in the aquatic environment, particularly in regard to factors such as aggregation (Cai et al, 2018). The natural environment is abundant in organic matter, and so the combined effects of organic matter and nanoplastics would need to be considered to fully understand the impact and fate of plastic nanoparticles in aquatic environments. Certainly, such factors would need to be considered when explaining trends in consumption, ingestion, and excretion of plastic nanoparticles by organisms in aquatic systems, but this is also complicated by variation in the shape, size, and chemistry of engineered nanoplastics now found in these environments (Cai et al, 2018). It has also been found that organic matter may promote bioaccumulation of and increase cell membrane damage by nanoplastics (Liang et al, 2020) but the mechanisms by which this may occur are unclear.

The mechanism or mode of interaction between nanoplastics and organic matter would also appear to differ depending on the nature of the organic matter, which complicates a clear understanding of these interactions (Liang et al, 2020; Wu et al, 2021). However, Chen et al. (2018) have observed accelerated assembly of dissolved organic matter into particulate organic matter in the presence of different types of nanoplastics, which indicates a potential influence of nanoplastics on the carbon component of aquatic environments (Chen et al, 2018).

3.2.4 Reactive Oxygen Species

The majority of studies pertaining to nanoplastics in the aquatic environment have focused on experimentally quantifying the toxicity of particular nanoplastic(s) of interest on a particular test organism or group of organisms. Many of the studies to date have considered the toxic effects of reactive oxygen species (ROS), such as the hydroxyl radical (OH·), superoxide radical ($O_2^{·-}$), and the non-radical reactive hydrogen peroxide molecule (H_2O_2), as a potential primary mechanism of toxicity (Bianco et al, 2020; Brandts et al, 2020).

There are many mechanisms (both endogenous and exogenous) which can result in production of ROS, just as there are many examples of ROS; but their common feature is their ease of production in water (in the case of exogenous production) and their potential toxicity arising from their highly reactive nature, which induces oxidative stress in the organism (Brandts et al, 2020). ROS production can be increased by UV exposure or elevated temperatures (Liu et al, 2020; Lu et al, 2017), which makes ROS toxicity potentially dependent on experimental conditions. Because ROS are generally reactive with many chemical compounds, they can induce a wide variety of negative effects in an organism, including cell membrane and DNA damage (Brandts et al, 2020; Feng et al, 2019; Liu et al, 2020).

Many studies have proposed toxicity through the generation of ROS by nanoplastics, but few have experimentally identified evidence for the importance of ROS-generating mechanisms, and fewer still challenging the limitations of our understanding regarding these processes. A strong example in favour of the ROS mechanism is the observation of cultured algal density increasing in the presence of polyhydroxy fullerenes (PHF) (Gao et al, 2011). Fullerenes are widely considered antioxidants, with strong radical scavenging properties, so it could be surmised that PHF scavenging ROS, thus reducing oxidative stress on algal communities. In the context of nanoplastics, Feng et al. (2019) saw increased growth inhibition of algae from polystyrene nanoplastics (PS-NPs) after inhibition of an intrinsic antioxidant response. Similarly, Liu et al. (2020) found over-production of ROS correlating to increasing concentrations of PS-NPs.

3.2.5 Formation and Degradation of Nanoplastics

While often resistant to weathering and break down due to their composition, plastics will certainly undergo some amount of degradation in aquatic environments (Mattsson et al, 2018), and degradation of plastic debris to 'secondary' microplastics has been examined closely in particular cases in the marine environment (Andrady, 2011). In the case of nanoplastics, there is a lot more research required on the ultimate source and fate of these particles. Determination of sources should include primary sources, such as engineered nanomaterials used in consumer products or industrial applications, and secondary sources, such as the degradation of macro or microplastics (Gonçalves & Bebianno, 2021). However, it is known that in some cases nanoplastics are known to be susceptible to degradation in the presence of ultra-violet (UV) light, where irradiation of PS-NPs has been observed to produce volatile compounds such as ketones, olefins, and styrene monomers, as well as degradation by mechanical processes (Piccardo et al, 2020). These processes may reduce the size of the nanoparticles, thus increasing their reactivity due to an increase in surface area, as discussed in section 1.2.1, but may also alter the surface charge (ZP) of the particle (Yu et al, 2019).

3.2.6 Nanoplastics as Vectors

Other chemical mechanisms beyond the production of ROS have been suggested to explain the toxicity of nanoplastics. One question that arises within the literature is that of the nano-specific toxicity hypothesis (Markiewicz et al, 2018). This hypothesis supposes that unusual properties of nano-scale materials are responsible for nano-specific mechanisms of toxicity. The counter-hypothesis is that the toxicity of nanomaterials arises from leaching of an ionic core metal (Fabrega et al, 2011).

An important distinction between nanoplastics and most other engineered nanomaterials is the lack of potential ion leachate.

Nanoplastics do not have a metal or metalloid core, so it is presumed that the sole methods of toxicity involve direct contact with the organism. However, while nanoplastics do not intrinsically contain metals or metalloids, additives are often added during the production process, and additional contaminants may be adsorbed to the nanoplastics in the natural environment. This makes the nanoplastic a potential vector for other contaminants, such as heavy metals (Baudrimont et al, 2020; Ferreira et al, 2019; Kumar et al, 2021).

The transport of pathogens by nanoplastics is a topic that has recently arisen in the literature (Wang et al, 2021). While a comprehensive understanding of the question is yet lacking, and little data has been proposed in support of the theory, tentative results have demonstrated that nanoplastics may aid in the migration of *Escherichia coli*, and there have been suggestions that nanoplastics may promote the translocation of bacteria and viruses (Wang et al, 2021).

3.2.7 Uptake and Accumulation

The uptake and accumulation of plastic pollution is the focus of many studies, particularly in marine environments, but increasingly in freshwater environments. While ingestion of meso- and micro-scale plastics by fish and marine mammals is widely observed and reported (Jacobsen et al, 2010; Lusher et al, 2015; Rummel et al, 2016), the extent to which nanoplastics are ingested by aquatic biota is difficult to determine. Accessible techniques employed in the identification of microplastics lack sufficient spatial resolution for the detection of nanoplastics (Ferreira et al, 2019).

In order to overcome these deficiencies, a number of studies have examined the retention of ingested nanoplastics in a laboratory setting. It should be noted that the concept of retention and bioaccumulation are frequently confused, and studies reporting the bioaccumulation of nanoplastics are often reporting retention, whereas the latter includes both indefinitely bioaccumulated quantities and quantities that will eventually be depurated. Additionally, a veritable mosaic of methodologies is employed, which hinders direct comparison of results. These shortcomings make it difficult to reconcile the findings of these studies. Increased scientific rigour, as well as standardization of terminology and approaches, is required in order to obtain a more accurate understanding of the uptake and accumulative behaviour of nanoplastics.

Despite these issues, there are clear indications that the persistence of nanoplastics in aquatic environments is an environmental threat that requires further careful investigation. An important barrier to toxins is the blood-brain barrier, whose correct operation is an essential neurotoxic defence. Of significant concern are studies that show evidence for traversal of the blood-brain barrier by nanoplastics (Kashiwada, 2006; Mattsson et al, 2017). The impacts of nanoplastics at higher trophic levels could include behavioural disorders (Chae & An, 2020; Mattsson et al, 2017), and trophic transfer of nanoplastics has been strongly suggested by a few multi-species studies (Cedervall et al, 2012; Chae & An, 2020; Mattsson et al, 2017); but there remains a significant lack of evidence for the importance of this pathway. An emphasis should be placed on improving our qualitative understanding of the effects of these nano-scale particles over mere quantification of their effects in environmentally irrelevant situations.

3.3 OUTLOOK

The recently increased interest in nanoplastics is already emerging as a new subfield of plastics pollution research. The specific properties and impacts of these analytes should be studied with the understanding that they are chemically distinct from metallic and metalloid nanomaterials, but equally different from micro- and meso-plastics, and findings and approaches related to the study of microplastics may not be widely transferable to the nanoplastic context (Gigault et al, 2021; Mitrano et al, 2021).

3.3.1 The Importance of Methodology

Nano-scale materials are not easily observed or quantified. Specialist analytical techniques and advanced microscopy allow us to characterize engineered nanoplastics *ex-situ*, but the identification and characterization of unknown environmental quantities remain beyond our current capabilities (Ferreira et al, 2019). Indeed, it has been previously pointed out that research on nanoplastics still suffers from a lack of commonly agreed protocols, and that there is a need for future studies to develop standardised techniques that allow for comparability of data (Horton et al, 2017). Common methods currently employed in the assessment of the ecological impact of contaminants focus on high-level impacts. This includes the quantification of the analyte in the environment and the evaluation of the toxicity of the analyte in an *ex-situ* context. In the case of nanoplastics, analytical limitations remove both options. Indeed, some approaches that are ideally suited to the evaluation of nanoparticle behaviours, such as analyte labelling (Mitrano et al, 2019), may interfere with the processes that the particle undergoes and produce environmentally unrealistic behaviour (Sander et al, 2019). For these reasons, methodological approaches that may have worked for larger particles are simply irrelevant in the case of nanoplastics (da Costa, 2018a; Gigault et al, 2021).

A more promising approach is to attempt to identify the mechanistic action of nanoplastics under given environmental conditions, with an aim to develop our qualitative understanding of possible modes of action, and quantify the relative importance of each. In essence, the scientist needs to move from the passive position of observing the distribution and ecotoxicological impact of the pollutant to engineering likely scenarios that can be interpreted and understood.

3.3.2 Recommendations

Nanoplastics are likely to undergo mechanisms unique to their size and composition. The distinction must be made between the toxicity of the nanoplastic itself, and the toxicity of potential additives and contaminants vectored by the plastic nanoparticle; if the identification of the role of non-polymeric substances is not ascertained, then it would be very easy to falsely assign toxic effects to the nanoplastic. We recommend that the potential toxicology of nanoplastic additives and other complexed contaminants be thoroughly examined, so as to not overestimate the risk of virgin nanoplastics.

It is evident in many studies that the focus of the field is not on the determination of mechanisms responsible for nanomaterial toxicity. The emphasis is usually placed on the determination of toxic risk rather than the qualitative behaviour of the material itself. Given the huge dimensionality of the problem at hand and difficulties in the quantification of nanoplastics, little progress can be made from simply assessing the toxicity of each different analyte to each class of organism under varying environmental conditions. A comprehensive understanding can only be attained through the determination of processes and mechanisms that can reliably be extrapolated, and an approach should be adopted that places more importance on the elucidation of *how* and *why* toxic effects occur, and less efforts be expended on quantifying toxicity unnecessarily.

Considerable benefit could be gained from read-across to other fields constrained by observational deficiencies and incorporation of more theoretical approaches. As there remain such gaps in our understanding of nanoplastics, and, not least, because the characterization of their environmental processes evades current techniques and comprehension, observation-based methodologies should perhaps give way to alternatives, such as molecular modelling and simulation. The question is, perhaps, better tackled from a materials science perspective than an ecotoxicological perspective, as the mechanisms of interaction cannot be directly observed and must be elucidated from a theoretical standpoint.

It has previously been pointed out that there has been little study of the impacts of nanoplastics on higher-level organisms, such as vertebrates. However, in these early days, this should not be taken as a criticism. From a mechanistic point of view, the popularity of microorganisms as test species is to be commended, as these simpler organisms will aid in the identification of processes at a cellular level. It is important to first develop a solid understanding of the mechanistic action of nanomaterials at a uni-cellular level before advancing to higher classes of organisms. Determination of the ecotoxicity of nanoplastics to more complex organisms can be hindered by a lack of understanding of the mechanisms and dependencies on environmental parameters and analyte characteristics; these are best approached from the bottom up.

3.4 CONCLUSIONS

Nanoplastics in aquatic environments (and elsewhere) are an emerging threat and are justifiably attracting increased scientific attention. It is important, however, that the distinction between nanoplastics and larger plastics is correctly appreciated. Nanoplastics require a different approach, appreciation of their nano-specific characteristics, and the employment of methodologies that do not assume the possibility for facile identification and quantification.

Future studies should endeavour to elucidate the likely processes and mechanism of toxicity associated with nanoplastics, and make no assumptions of possible transferability of previous meso- and micro-scale results.

REFERENCES

Almutairi, M. M. A., Gong, C., Xu, Y. G., Chang, Y. & Shi, H. (2016) Factors controlling permeability of the blood–brain barrier. *Cellular and Molecular Life Sciences* 73, 57–77. 10.1007/s00018-015-2050-8.

Andrady, A. L. (2011) Microplastics in the marine environment. *Marine Pollution Bulletin* 62, 1596–1605. 10.1016/j.marpolbul.2 011.05.030.

Barnes, D. K. A., Galgani, F., Thompson, R. C. & Barlaz, M. (2009) Accumulation and fragmentation of plastic debris in global environments. *Philosophical Transactions of the Royal Society B* 364, 1985–1998. 10.1098/rstb.2008.0205.

Baudrimont, M., Arini, A., Guégan, C., Venel, Z., Gigault, J., Pedrono, B., Prunier, J., Maurice, L., Ter Halle, A. & Feurtet-Mazel, A. (2020) Ecotoxicity of polyethylene nanoplastics from the North Atlantic oceanic gyre on freshwater and marine organisms (microalgae and filter-feeding bivalves). *Environmental Science and Pollution Research* 27, 3746–3755. 10.1007/s11356-019-04668-3.

Bianco, A., Sordello, F., Ehn, M., Vione, D. & Passananti, M. (2020) Degradation of nanoplastics in the environment: Reactivity and impact on atmospheric and surface waters. *Science of The Total Environment* 742, 140413. 10.1016/j.scitotenv.2020.140413.

Boyle, K. & Örmeci, B. (2020) Microplastics and nanoplastics in the freshwater and terrestrial environment: A review. *Water* 12, 2633. 10.3390/w12092633.

Brandts, I., Garcia-Ordoñez, M., Tort, L., Teles, M. & Roher, N. (2020) Polystyrene nanoplastics accumulate in ZFL cell lysosomes and in zebrafish larvae after acute exposure, inducing a synergistic immune response in vitro without affecting larval survival *in vivo*. *Environmental Science: Nano* 7, 2410–2422. 10.1039/D0EN00553C.

Cai, L., Hu, L., Shi, H., Ye, J., Zhang, Y. & Kim, H. (2018) Effects of inorganic ions and natural organic matter on the aggregation of nanoplastics. *Chemosphere* 197, 142–151. 10.1016/j.chemosphere.201 8.01.052.

Cedervall, T., Hansson, L.-A., Lard, M., Frohm, B. & Linse, S. (2012) Food chain transport of nanoparticles affects behaviour and fat metabolism in fish. *PLoS ONE* 7, e32254. 10.1371/journal.pone.0032254.

Chae, Y. & An, Y.-J. (2020) Nanoplastic ingestion induces behavioral disorders in terrestrial snails: trophic transfer effects via vascular plants. *Environ. Sci.: Nano* 7, 975–983. 10.1039/C9EN01335K.

Chen, C.-S., Le, C., Chiu, M.-H. & Chin, W.-C. (2018) The impact of nanoplastics on marine dissolved organic matter assembly. *Science of The Total Environment* 634, 316–320. 10.1016/j.scitotenv.2018.03.269.

da Costa, J. P. (2018a) Micro- and nanoplastics in the environment: Research and policymaking. *Current Opinior. in Environmental Science & Health* 1, 12–16. 10.1016/j.coesh.2017.11.002.

da Costa, J. P. (2018b) Nanoplastics in the environment, in Harrison, R. M. and Hester, R. E. eds. *Issues in Environmental Science and Technology.* Cambridge: Royal Society of Chemistry, pp. 82–105. 10.1039/9781788013314-00082.

Donaldson, K. & Poland, C. A. (2013) Nanotoxicity: challenging the myth of nano-specific toxicity. *Current Opinion in Biotechnology* 24, 724–734. 10.1016/j.copbio.2013.05.003.

Fabrega, J., Luoma, S. N., Tyler, C. R., Galloway, T. S. & Lead, J. R. (2011) Silver nanoparticles: Behaviour and effects in the aquatic environment. *Environment International* 37, 517–531. 10.1016/j.envint.2010.10.012.

Feng, L.-J., Li, J.-W., Xu, E. G., Sun, X.-D., Zhu, F.-P., Ding, Z., Tian, H., Dong, S.-S., Xia, P.-F. & Yuan, X.-Z. (2019) Short-term exposure to positively charged polystyrene nanoparticles causes oxidative stress and membrane destruction in cyanobacteria. *Environmental Science: Nano* 6, 3072–3079. 10.1039/C9EN00807A.

Ferreira, I., Venâncio, C., Lopes, I. & Oliveira, M. (2019) Nanoplastics and marine organisms: What has been studied? *Environmental Toxicology and Pharmacology* 67, 1–7. 10.1016/j.etap.2019.01.006.

Gao, J., Wang, Y., Folta, K. M., Krishna, V., Bai, W., Indeglia, P., Georgieva, A., Nakamura, H., Koopman, B. & Moudgil, B. (2011) Polyhydroxy fullerenes (Fullerols or Fullerenols): Beneficial effects on growth and lifespan in diverse biological models. *PLoS ONE* 6, e19976. 10.1371/journal.pone.0019976.

Gigault, J., El Hadri, H., Nguyen, B., Grassl, B., Rowenczyk, L., Tufenkji, N., Feng, S. & Wiesner, M. (2021) Nanoplastics are neither microplastics nor engineered nanoparticles. *Nature Nanotechnology* 16, 501–507. 10.1038/s41565-021-00886-4.

Gonçalves, J. M. & Bebianno, M. J. (2021) Nanoplastics impact on marine biota: A review. *Environmental Pollution* 273, 116426. 10.1016/j.envpol.2021.116426.

González-Pleiter, M., Tamayo-Belda, M., Pulido-Reyes, G., Amariei, G., Leganés, F., Rosal, R. & Fernández-Piñas, F. (2019) Secondary nanoplastics released from a biodegradable microplastic severely impact freshwater environments. *Environmental Science: Nano* 6, 1382–1392. 10.1039/C8EN01427B.

Hartmann, N. B., Hüffer, T., Thompson, R. C., Hassellöv, M., Verschoor, A., Daugaard, A. E., Rist, S., Karlsson, T., Brennholt, N., Cole, M., Herrling, M. P., Hess, M. C., Ivleva, N. P., Lusher, A. L. & Wagner, M. (2019) Are we speaking the same language? Recommendations for a definition and categorization framework for plastic debris. *Environmental Science & Technology* 53, 1039–1047. 10.1021/acs.est.8b05297.

Horton, A. A., Walton, A., Spurgeon, D. J., Lahive, E. & Svendsen, C. (2017) Microplastics in freshwater and terrestrial environments: Evaluating the current understanding to identify the knowledge gaps and future research priorities. *Science of The Total Environment* 586, 127–141. 10.1016/j.scitotenv.2017.01.190.

Huang, D., Tao, J., Cheng, M., Deng, R., Chen, S., Yin, L. & Li, R. (2021) Microplastics and nanoplastics in the environment: Macroscopic transport and effects on creatures. *Journal of Hazardous Materials* 407, 124399. 10.1016/j.jhazmat.2020.124399.

Jacobsen, J. K., Massey, L. & Gulland, F. (2010) Fatal ingestion of floating net debris by two sperm whales (*Physeter macrocephalus*). *Marine Pollution Bulletin* 60, 765–767. 10.1016/j.marpolbul.201 0.03.008.

Kashiwada, S. (2006) Distribution of nanoparticles in the see-through Medaka (*Oryzias latipes*). *Environmental Health Perspectives* 114, 1697–1702. 10.1289/ehp.9209.

Khoshnamvand, M., Hanachi, P., Ashtiani, S. & Walker, T. R. (2021) Toxic effects of polystyrene nanoplastics on microalgae *Chlorella vulgaris*: Changes in biomass, photosynthetic pigments and morphology. *Chemosphere* 280, 130725. 10.1016/j.chemosphere.2 021.130725.

Koelmans, A. A., Besseling, E. & Shim, W. J. (2015) Nanoplastics in the aquatic environment. *Critical Review*, in Bergmann, M., Gutow, L. and Klages, M. eds. *Marine Anthropogenic Litter*. Cham: Springer International Publishing, pp. 325–340. 10.1007/978-3-319-1651 0-3_12.

Kukkola, A., Krause, S., Lynch, I., Sambrook Smith, G. H. & Nel, H. (2021) Nano and microplastic interactions with freshwater biota – Current knowledge, challenges and future solutions. *Environment International* 152, 106504. 10.1016/j.envint.2021.106504.

Kumar, M., Chen, H., Sarsaiya, S., Qin, S., Liu, H., Awasthi, M. K., Kumar, S., Singh, L., Zhang, Z., Bolan, N. S., Pandey, A., Varjani, S. & Taherzadeh, M. J. (2021) Current research trends on micro-

and nano-plastics as an emerging threat to global environment: A review. *Journal of Hazardous Materials* 409, 124967. 10.1016/j.jhazmat.2020.124967.

Liang, D., Wang, X., Liu, S., Zhu, Y., Wang, Y., Fan, W. & Dong, Z. (2020) Factors determining the toxicity of engineered nanomaterials to *Tetrahymena thermophila* in freshwater: The critical role of organic matter. *Environmental Science: Nano* 7, 304–316. 10.1039/C9EN01017C.

Liu, Y., Wang, Z., Wang, S., Fang, H., Ye, N. & Wang, D. (2019) Ecotoxicological effects on *Scenedesmus obliquus* and *Danio rerio* Co-exposed to polystyrene nano-plastic particles and natural acidic organic polymer. *Environmental Toxicology and Pharmacology* 67, 21–28. 10.1016/j.etap.2019.01.007.

Liu, Z., Huang, Y., Jiao, Y., Chen, Q., Wu, D., Yu, P., Li, Y., Cai, M. & Zhao, Y. (2020) Polystyrene nanoplastic induces ROS production and affects the MAPK-HIF-1/NFkB-mediated antioxidant system in *Daphnia pulex*. *Aquatic Toxicology* 220, 105420. 10.1016/j.aquatox.2020.105420.

Lu, H., Fan, W., Dong, H. & Liu, L. (2017) Dependence of the irradiation conditions and crystalline phases of TiO_2 nanoparticles on their toxicity to *Daphnia magna*. *Environmental Science: Nano* 4, 406–414. 10.1039/C6EN00391E.

Lusher, A. L., Hernandez-Milian, G., O'Brien, J., Berrow, S., O'Connor, I. & Officer, R. (2015) Microplastic and macroplastic ingestion by a deep diving, oceanic cetacean: the True's beaked whale *Mesoplodon mirus*. *Environmental Pollution* 199, 185–191. 10.1016/j.envpol.2015.01.023.

Markiewicz, M., Kumirska, J., Lynch, I., Matzke, M., Köser, J., Bemowsky, S., Docter, D., Stauber, R., Westmeier, D. & Stolte, S. (2018) Changing environments and biomolecule coronas: consequences and challenges for the design of environmentally acceptable engineered nanoparticles. *Green Chemistry* 20, 4133–4168. 10.1039/C8GC01171K.

Mateos-Cárdenas, A., O'Halloran, J., van Pelt, F. N. A. M. & Jansen, M. A. K. (2020) Rapid fragmentation of microplastics by the freshwater amphipod *Gammarus duebeni* (Lillj.). *Scientific Reports* 10, 12799. 10.1038/s41598-020-69635-2.

Mattsson, K., Jocic, S., Doverbratt, I. & Hansson, L.-A. (2018) Nanoplastics in the aquatic environment, in *Microplastic*

Contamination in Aquatic Environments. Elsevier, pp. 379–399. 10.1016/B978-0-12-813747-5.00013-8.

Mattsson, K., Johnson, E. V., Malmendal, A., Linse, S., Hansson, L.-A. & Cedervall, T. (2017) Brain damage and behavioural disorders in fish induced by plastic nanoparticles delivered through the food chain. *Scientific Reports* 7, 11452. 10.1038/s41598-017-10813-0.

Mitrano, D. M., Beltzung, A., Frehland, S., Schmiedgruber, M., Cingolani, A. & Schmidt, F. (2019) Synthesis of metal-doped nanoplastics and their utility to investigate fate and behaviour in complex environmental systems. *Nature Nanotechnology* 14, 362–368. 10.1038/s41565-018-0360-3.

Mitrano, D. M., Wick, P. & Nowack, B. (2021) Placing nanoplastics in the context of global plastic pollution. *Nature Nanotechnology* 16, 491–500. 10.1038/s41565-021-00888-2.

Munk, M., Brandão, H. M., Yéprémian, C., Couté, A., Ladeira, L. O., Raposo, N. R. B. & Brayner, R. (2017) Effect of multi-walled carbon nanotubes on metabolism and morphology of filamentous green microalgae. *Archives of Environmental Contamination and Toxicology* 73, 649–658. 10.1007/s00244-017-0429-2.

Oriekhova, O. & Stoll, S. (2018) Heteroaggregation of nanoplastic particles in the presence of inorganic colloids and natural organic matter. *Environmental Science: Nano* 5, 792–799. 10.1039/C7EN01119A.

Piccardo, M., Renzi, M. & Terlizzi, A. (2020) Nanoplastics in the oceans: Theory, experimental evidence and real world. *Marine Pollution Bulletin* 157, 111317. 10.1016/j.marpolbul.2020.111317.

Rummel, C. D., Löder, M. G. J., Fricke, N. F., Lang, T., Griebeler, E.-M., Janke, M. & Gerdts, G. (2016) Plastic ingestion by pelagic and demersal fish from the North Sea and Baltic Sea. *Marine Pollution Bulletin* 102, 134–141. 10.1016/j.marpolbul.2015.11.043.

Sander, M., Kohler, H.-P. E. & McNeill, K. (2019) Assessing the environmental transformation of nanoplastic through 13C-labelled polymers. *Nature Nanotechnology* 14, 301–303. 10.1038/s41565-019-0420-3.

Shen, M., Zhang, Y., Zhu, Y., Song, B., Zeng, G., Hu, D., Wen, X. & Ren, X. (2019) Recent advances in toxicological research of nanoplastics in the environment: A review. *Environmental Pollution* 252, 511–521. 10.1016/j.envpol.2019.05.102.

Strungaru, S.-A., Jijie, R., Nicoara, M., Plavan, G. & Faggio, C. (2019) Micro- (nano) plastics in freshwater ecosystems: Abundance, toxicological impact and quantification methodology. *TrAC Trends in Analytical Chemistry* 110, 116–128. 10.1016/j.trac.201 8.10.025.

Sun, X.-D., Yuan, X.-Z., Jia, Y., Feng, L.-J., Zhu, F.-P., Dong, S.-S., Liu, J., Kong, X., Tian, H., Duan, J.-L., Ding, Z., Wang, S.-G. & Xing, B. (2020) Differentially charged nanoplastics demonstrate distinct accumulation in *Arabidopsis thaliana*. *Nature Nanotechnology*. 15, 755–760. 10.1038/s41565-020-0707-4.

Wagner, S. & Reemtsma, T. (2019) Things we know and don't know about nanoplastic in the environment. *Nature Nanotechnology* 14, 300–301. 10.1038/s41565-019-0424-z.

Wang, L., Wu, W.-M., Bolan, N. S., Tsang, D. C. W., Li, Y., Qin, M. & Hou, D. (2021) Environmental fate, toxicity and risk management strategies of nanoplastics in the environment: Current status and future perspectives. *Journal of Hazardous Materials* 401, 123415. 10.1016/j.jhazmat.2020.123415.

Wu, J., Jiang, R., Liu, Q. & Ouyang, G. (2021) Impact of different modes of adsorption of natural organic matter on the environmental fate of nanoplastics. *Chemosphere* 263, 127967. 10.1016/j.chemosphere.2020.127967.

Yu, F., Yang, C., Zhu, Z., Bai, X. & Ma, J. (2019) Adsorption behavior of organic pollutants and metals on micro/nanoplastics in the aquatic environment. *Science of The Total Environment* 694, 133643. 10.1016/j.scitotenv.2019.133643.

CHAPTER 4

Considerations for the Pharmaceutical Industry Regarding Environmental and Human Health Impacts of Microplastics

Yvonne Lang[1], Sandra O'Neill[2], and Jenny Lawler[2,3]

[1]*Centre for Environmental Research, Innovation and Sustainability (CERIS), Institute of Technology Sligo, Ash Lane, SligoF91 YW50, Ireland*
[2]*School of Biotechnology and DCU Water Institute, Dublin City University, DCU Glasnevin Campus, Dublin 9, Ireland*
[3]*Qatar Environment and Energy Research Institute, Hamad Bin Khalifa University, Doha, Qatar*

DOI: 10.1201/9781003109730-4

CONTENTS

4.1 INTRODUCTION

Microplastics (MPs) are one of the most ubiquitous pollutants released into the environment (Ali et al, 2021; De-la-Torre, 2020). Primary sources of MPs are particles intentionally manufactured at this scale. Secondary sources are derived from the breakdown of larger plastics. Global production of plastics was 370 million tonnes in 2019, with 58 million tonnes produced in Europe (PlasticsEurope, 2020). The breakdown use of plastics is typically reported by end-use with packaging representing the largest share in the European market at 39.6%. A detailed analysis of the scale of plastics and microplastics either used or produced in the pharmaceutical industry is not available. An evaluation of the full product lifecycle of a pharmaceutical product is required to gain an appreciation of the scale of plastic use in the pharmaceutical industry. This process will identify plastic used during the manufacturing process related to protective equipment, materials, equipment (including single-use equipment that is common in the biopharmaceutical industry), sampling, testing, storage, and packaging. Understanding the volume of all plastics used in the lifecycle and disposal pathways will assist in truly quantifying the role of the pharmaceutical industry on environmental and human exposure to microplastics.

It is estimated that single-use plastics generated by the bio-pharmaceutical industry contribute 30,000 tonnes of plastic waste per year to landfill or incineration (McDonald, G.J., 2019). Furthermore, the continued annual growth rate of the pharmaceutical plastic packaging market is forecast at 8.24% between the years 2021–2026 (Research & Markets, 2021). In Canada, 250 million drug prescriptions are dispensed in packaging that contributes 6,000 tonnes of plastic waste annually (EcoloPharm, 2019). MPs can be added to drug formulations as an excipient, defined as an inactive substance that serves as the vehicle for an active pharmaceutical ingredient. The pharmaceutical industry has undoubtedly improved healthcare worldwide (global life expectancy has increased from 66.8 years to 73.4 years between 2000 and 2019 (World Health Organisation, 2021). Consumption of pharmaceuticals for treatment of aging-related diseases in some cases quadrupled in OECD countries between 2000 and 2015 (González Peña et al, 2021). As such, it is somewhat of a paradox that this industry potentially contributes to the adverse impacts on environmental health and human health through the possible release of MPs into the environment via degradation of plastic items such as solution containers, transfusion sets, transfer tubing, drug packaging, and devices.

Pharmaceuticals have been widely detected in all parts of the environment and it is acknowledged that there are three mechanisms by which this occurs: domestic wastewater, incorrectly disposed pharmaceuticals, and industrial related release (aus der Beek et al, 2016; European Commission, 2019; Khan et al, 2021). It is suggested that pharmaceuticals can interact with MPs present in the environment, and alter the chemical properties of MPs leading to an enhancement of possible toxic effects (Zhou et al, 2020). The scale of this problem is not fully known nor are the potential environmental and health impacts.

The role of the pharmaceutical industry related to MPs is therefore multifaceted; what is the scale of primary MPs use in pharmaceutical products?; what is the scale of plastic use in the pharmaceutical industry that could contribute to the generation of secondary MPs?; what is the impact of interaction between pharmaceuticals and MPs in the environment? This chapter will explore these issues and highlight actions occurring at European level to address this problem.

4.2 PHARMACEUTICAL INDUSTRY AND THE GENERATION OF PLASTIC WASTE

In the USA, healthcare facilities generate on average 14,000 tons of waste per day where a quarter is made up of sterile plastic products such as packaging (HPRC, 2020). The industry is well regulated for leachates and extractables from both product contact materials used in manufacture, and packaging materials (Jenke, 2015a, b; 2021); however, breakdown of the plastic products themselves after consumption of the pharmaceutical is not widely considered. The pharmaceutical packaging market is expected to exceed 84 billion GBP by 2024 (Origin Pharma Packaging, 2021), consisting of primary packaging (blister packs, bottles, aerosols), secondary packaging (cartons and boxes), and tertiary packaging (barrels, edge protectors, and containers). The increase in the market is driven by factors such as increasing demand for prescription dose medicines, growing numbers of patients with chronic diseases, an increase in accessibility to medicines in developing markets such as India, an increasing prevalence of generics in the market, and the growing demand for environmentally friendly packaging (Transparency Market Research, 2021). The industry acknowledges the need to move towards more sustainable practices regarding packaging, and there are expected changes on the horizon to move towards bioplastics, biodegradable plastics, and compostable plastics as a result.

4.3 THE INTERACTION OF PHARMACEUTICALS WITH MPS

There is substantial evidence demonstrating the presence of pharmaceuticals in the environment. An extensive review of the literature was published in 2016 by Küster and colleagues outlining the global occurrence of pharmaceuticals presenting very concerning results regarding the scale of this issue (aus der Beek et al, 2016). Some of the key findings of the review were that 631 pharmaceutical substances were detected in the environment in 71 countries and the main pathways by which pharmaceuticals make their way into the environment is via urban wastewater. The review included publications up to and including October 2013 and further research in the interim has detected additional pharmaceutical products in the environment (Maculewicz et al, 2022; Patel et al, 2019). A key issue for consideration regarding environmental and human health impacts is the interaction of pharmaceutical residues and transformation products with the vast quantities of MPs in the environment from all sources.

The first question to address is in relation to the quantities of pharmaceuticals present in the environment and whether concentrations have reached a threshold of concern. It is acknowledged that the levels of pharmaceuticals in certain matrices such as drinking water are not of immediate concern. However, it is prudent to maintain the viewpoint that exposure to these levels over a sustained period of time has the potential to have adverse effects on the body. It is unknown what impacts chronic exposure to persistent entities in the environment will have. These entities will also be increasing in quantity with time. Therefore, the viewpoint must be that an unsafe threshold will be reached if action is not taken.

Investigating the sorption of pharmaceuticals onto microplastics is an area of great importance. Recent review articles provide a very thorough overview of the state of knowledge of pharmaceutical-microplastic interactions from the context of sorption (Atugoda et al, 2021; Vieira et al, 2021). In summary, adsorption capacities are dependent on the pharmaceutical, microplastic composition, age of the microplastic, and environmental conditions such as ionic strength, temperature, pH, and presence of dissolved organic matter (Puckowski et al, 2021; Wu et al, 2016; Xu et al, 2021). There is evidence that the toxicity of a pharmaceutical compound may be greater when it is adsorbed on the surface of a microplastic (Fonte et al, 2016; Prata et al, 2018; Qu et al, 2019; Zhang et al, 2019).

A significant body of work is required to fully understand the potential toxicity of pharmaceutical-microplastic interactions. Many studies use pristine plastics or plastic powders when evaluating the interaction of microplastics and pharmaceuticals (Elizalde-Velázquez et al, 2020; Puckowski et al, 2021). While it is vitally important to conduct preliminary studies with the material in isolation follow-up studies are required to demonstrate that the pharmaceutical will interact with real-world samples. The number of confounding factors that will influence this analysis is acknowledged and making correlations will be difficult. Extrapolating results from pristine samples to the real world will have significant hurdles also and may ultimately over-estimate the scale of sorption and thus potential health impact.

Should greater importance be placed on the evaluation of drug metabolites and waste products onto microplastics? It has been suggested that the toxicity of pharmaceuticals to aquatic organisms is not necessarily from the parent compound but that there is also a role for degradants and transformation products (Maculewicz et al, 2022; Prata, 2018; Prata et al, 2018).

Are other components of the environmental matrix going to influence the toxicity of the pharmaceutical-microplastic "composite"? Environmental samples are complex and compounds other than pharmaceuticals may compete for the microplastic. The competition that exists may mean that another compound is adsorbed preferentially. As stated previously a number of factors have been shown to influence adsorption of pharmaceuticals.

The interactions are so complex that identifying a single element of the chain as a main contributor is somewhat impossible.

The interaction of pharmaceuticals with biofilms on microplastics is also under investigation. It has been suggested that the presence of dissolved organic matter can interfere with the sorption process through competition for adsorption/absorption sites. Microorganisms may respond to the presence of microplastics by potentially exuding extracellular polymeric substances (EPS). This EPS may potentially coat the plastic particle and alter the surface chemistry and thus sorption. The EPS may also enable attachment of the microorganism to the microplastic. Two additional points need to be considered when evaluating this pharmaceutical-microplastic "composite" (a) the microplastic may have pharmaceuticals already sorbed previous to EPS coating (b) microorganisms, such as microalgae, have been shown to sequester pharmaceuticals from the environment. Therefore, the following scenario may occur, where a microorganism reacts to the presence of microplastics in the environment leading to an increased production of EPS that attaches to the microplastic, thus creating enhanced pharmaceutical sorption. The microorganism that has already sequestered a pharmaceutical also attaches to the particle. Thus, there is double concentration-effect – pharmaceutical present in the microorganism and pharmaceutical present on the microplastic. The final concentration of pharmaceutical may exceed safe thresholds.

The importance of reliable methodologies for quantifying pharmaceuticals in real-world samples is paramount. Laboratory experiments are conducted under very controlled settings and thus evaluation of pharmaceutical-microplastic interactions, while not a simple task, is easier than for real-world samples. Recent research describes methodologies for the extraction and quantitative determination of ten pharmaceuticals from real-world microplastic (Santana-Viera et al, 2021). Additional work on developing robust methodologies and standard protocols is required.

It is accepted that pharmaceuticals enter the environment via three pathways; domestic wastewater, incorrectly disposed pharmaceuticals, and industrial related release (aus der Beek et al, 2016; European Commission, 2019; Khan et al, 2021). With changing legislation for environmental presence of pharmaceuticals, there is a strong onus to develop wastewater treatment systems to adequately remove pharmaceutical residues (European Commission, 2020a). Pharmaceuticals are often not well removed in typical secondary wastewater treatment systems (Matamoros et al, 2009; Nguyen et al, 2021; Verlicchi et al, 2013). As such, despite the current lack of legislation globally on discharge limits for pharmaceuticals and pharmaceutical and personal care products from wastewater treatment plants, a variety of technological advances targeting this complex matrix have been developed, often in the form of tertiary treatments. These include processes such as membrane bioreactors (Femina Carolin et al, 2021), advanced oxidation processes (Krishnan et al, 2021), and enzyme-based processes (Hena et al, 2021) (which are advantageous as breakdown of the compounds takes place), sorption processes (which require subsequent treatment and regeneration of the sorbent) (Fallah et al, 2021), and electrocoagulation processes (which require sludge disposal) (Zaied et al, 2020).

Correct disposal schemes need to be introduced in tandem with education and awareness campaigns. There is an ad-hoc approach to disposal schemes and this can make it prohibitive for the end-user to dispose of unused pharmaceutical products in a responsible manner.

4.4 PHARMACEUTICALS IN THE ENVIRONMENT AND CURRENT EU POLICIES TO ADDRESS THE ISSUE

The importance of addressing the problem of pharmaceuticals in the environment is acknowledged by the European Union (EU). The EU Strategic Policy on Pharmaceuticals in the Environment was adopted in March 2019 outlining approaches to address the risks posed to environmental and human health by human and veterinary pharmaceuticals throughout the full lifecycle of the product. The report outlines six areas for action: (1) Increase awareness and promote prudent use of pharmaceuticals; (2) Support the development of pharmaceuticals intrinsically less harmful for the environment and promote greener manufacturing; (3) Improve environmental risk assessment and its review; (4) Reduce wastage and improve the management of waste; (5) Expand environmental monitoring; (5) Fill other knowledge gaps. The report outlines 28 actions within these six areas involving engagement with stakeholders across public bodies, research bodies, the pharmaceutical industry, healthcare, and medical professionals, and the public. Furthermore, the actions emphasize the need to link to other relevant EU strategies and directives such that work is not duplicated. The following directives, initiatives, and research projects are mentioned in the actions: the Water Framework Directive, the Industrial Emissions Directive, Innovative Medicines Initiative, REACH, Information Platform for Chemical Monitoring, and European Commission's LUCAS soil survey. In addition, it was pointed out that there is a need for sharing of information and co-operation both within the EU (through the member states and various agencies) as well as with agencies that have reach outside of the EU such as the World Health Organisation.

The Pharmaceutical Strategy for Europe (European Commission, 2020b) addresses the importance of human and ecosystem health concerned with pharmaceuticals in the environment. The topic of environmentally sustainable pharmaceuticals is discussed in the document and it is acknowledged that *"Action is required throughout the lifecycle of medicines to reduce resource use, emissions and levels of pharmaceutical residues in the environment."* The strategy proposes two flagship initiatives regarding pharmaceutical legislation whereby proposals will be submitted to ensure overall environmental sustainability and strengthen the environmental risk assessments included in the legislation. Other actions involve working with

Member States and stakeholders in such areas as good manufacturing practices, information sharing, and decarbonizing value
chains.

The pharmaceutical industry is actively involved in the cam-
paign to reduce pharmaceuticals in the environment through the
Eco-Pharmaco-Stewardship (EPS) initiative (EFPIA, 2015). This
approach takes account of the entire lifecycle of the pharmaceutical product and while the industry acknowledges and ac-
cepts it has a very important role and responsibility in reducing
pharmaceuticals in the environment it also points out roles for
other entities including public services, environmental experts,
doctors, pharmacists, and patients. Healthcare professionals are
actively involved in awareness campaigns for the public such as
Safe Pharma (HCWH, 2021).

4.5 PHARMACEUTICALS AND INTENTIONALLY ADDED MICROPLASTICS

The Pharmaceutical Industry may face many challenges if the
proposed ban on intentionally added microplastics is brought
forward. The ECHA have provided a definition of microplastic
for regulatory purposes in the proposal as follows:

> "material consisting of solid polymer-containing parti-
> cles, to which additives or other substances may have
> been added, and where $\geq 1\%$ weight by weight (w/w) of
> particles have (i) all dimensions $1\ nm \leq x \leq 5\ mm$, or
> (ii), for fibers, a length of $3\ nm \leq x \leq 15\ mm$ and length
> to diameter ratio of >3. Polymers that occur in nature
> that have not been chemically modified (other than by
> hydrolysis) are excluded, as are polymers that are (bio)
> degradable"

There is a derogation for medicinal products that have been
welcomed by the pharmaceutical industry (EFPIA, 2019).
Excipients can commonly include polymers such as cellulose

acetate, hydroxypropylcellulose, polyvinylpolypyrrolidone, hydroxypropylmethylcellulose phthalate, polymethacrylates, polyethylenglycol, and microcrystalline cellulose. Many of these excipients are approved safe for use according to the European Pharmacopoeia. Sourcing an alternative to polymer excipients may be an arduous task as alternatives are very limited. The replacement of an excipient is not a simple task and may require many years of research to achieve a formulation that will be as efficacious, stable, and safe as the original formulation. However, there is a request to report used excipients on an annual basis and labelling.

4.6 QUANTIFYING EXPOSURE TO MICROPLASTICS VIA INGESTION OF PHARMACEUTICAL PRODUCTS

Pharmaceutical products for oral, parenteral, topical, and rectal administration include MPs as an excipient (Debotton & Dahan, 2017). Some understanding has been gained on the distribution of MPs following oral ingestions (O'Neill & Lawler, 2021). However, there is a paucity of studies examining the extent of exposure to MPs that may occur via other routes. The presence of MPs in food and water has been reported extensively providing information that will be of use in understanding the factors that affect distribution (Akoueson et al, 2020; Barboza et al, 2020; Conti et al, 2020; Shruti et al, 2020; Zuccarello et al, 2019). It has been reported that uptake and translocation are influenced by particle size, surface charge, hydrophobicity, and surface functionalization (Stock et al, 2019; Stock et al, 2020; Stock et al, 2021; Wu et al, 2019). Advances in drug development have involved the use of polymeric materials as drug coatings, excipients, and capsules for controlled release. Examples of such materials include cellulose acetate, hydrxpropylcellulose, polyvinylpyrrolidone, and polymethacrylates. Extensive research is ongoing in the field of polymeric capsules for targeted therapeutics.

As this research continues a number of questions must be answered regarding the volume of MPs generated and practices required to prevent further release of MPs into the environment. For example, (1) are there regulations in place for the correct disposal of polymeric waste during the research stage across public and private bodies? (2) what is the degradation profile of these materials in the body? (3) what is the degradation profile of these materials in the environment? It is vitally important that the research community move away from the practice of retrospective corrective actions, and instead implement and embed protective actions in research activities.

4.7 CONCLUSIONS

The pharmaceutical industry has contributed to improved human health. Treatment of diseases that were once fatal now exists. Life expectancy has increased. While the benefits derived from this sector are undisputed, there is in parallel increasing evidence indicating an adverse impact on environmental health from the pharmaceutical industry. This impact may be direct and/or indirect, that is, release of material into the environment directly by the industry, or release of material by other pathways such as waste products generated by the end-user. There is no doubt that there are multiple players involved in how pharmaceutical products may find their way into the environment, and thus roles and responsibilities exist for the various stakeholders. It is important that all players take ownership of their contribution to the problem and commence actions to mitigate and remediate their potential impact.

REFERENCES

Akoueson, F., Sheldon, L. M., Danopoulos, E., Morris, S., Hotten, J., Chapman, E., Li, J. & Rotchell, J. M. (2020) A preliminary analysis of microplastics in edible versus non-edible tissues from seafood samples. *Environmental Pollution*, 263, 114452.

Ali, M. U., Lin, S., Yousaf, B., Abbas, Q., Munir, M. A. M., Ali, M. U., Rasihd, A., Zheng, C., Kuang, X. & Wong, M. H. (2021) Environmental emission, fate and transformation of microplastics in biotic and abiotic compartments: Global status, recent advances and future perspectives. *Science of The Total Environment*, 791, 148422.

Atugoda, T., Vithanage, M., Wijesekara, H., Bolan, N., Sarmah, A. K., Bank, M. S., You, S. & Ok, Y. S. (2021) Interactions between microplastics, pharmaceuticals and personal care products: Implications for vector transport. *Environ Int*, 149, 106367.

aus der Beek, T., Weber, F.-A., Bergmann, A., Hickmann, S., Ebert, I., Hein, A. & Küster, A. (2016) Pharmaceuticals in the environment—Global occurrences and perspectives. *Environmental Toxicology and Chemistry*, 35(4), 823–835.

Barboza, L. G. A., Lopes, C., Oliveira, P., Bessa, F., Otero, V., Henriques, B., Raimundo, J., Caetano, M., Vale, C. & Guilhermino, L. (2020) Microplastics in wild fish from North East Atlantic Ocean and its potential for causing neurotoxic effects, lipid oxidative damage, and human health risks associated with ingestion exposure. *Science of The Total Environment*, 717, 134625.

Conti, G. O., Ferrante, M., Banni, M., Favara, C., Nicolosi, I., Cristaldi, A., Fiore, M. & Zuccarello, P. (2020) Micro- and nano-plastics in edible fruit and vegetables. The first diet risks assessment for the general population. *Environmental Research*, 187.

De-la-Torre, G. E. (2020) Microplastics: An emerging threat to food security and human health. *Journal of food science and technology*, 57(5), 1601–1608.

Debotton, N. & Dahan, A. (2017) Applications of polymers as pharmaceutical excipients in solid oral dosage forms. *Medicinal Research Reviews*, 37(1), 52–97.

EcoloPharm (2019) *Reducing Waste in Pharmacies*, 2019. Available online: https://www.ecolopharm.com/en/think-green/plastic-pollution-pharmacy/.

EFPIA (2015) *Eco-Pharmaco-Stewardship (EPS) - A holistic environmental risk management program*, 2015. Available online: https://www.efpia.eu/news-events/the-efpia-view/efpia-news/151009-eco-pharmaco-stewardship-eps-a-holistic-environmental-risk-management-program/.

EFPIA (2019) *EFPIA position on the ECHA Microplastics restrictions*, 2019. Available online: https://www.efpia.eu/media/554625/efpia-position-echa-microplastics-restrictions.pdf.

Elizalde-Velázquez, A., Subbiah, S., Anderson, T. A., Green, M. J., Zhao, X. & Cañas-Carrell, J. E. (2020) Sorption of three common nonsteroidal anti-inflammatory drugs (NSAIDs) to microplastics. *Sci Total Environ*, 715, 136974.

European Commission (2019) COMMUNICATION FROM THE COMMISSION TO THE EUROPEAN PARLIAMENT, THE COUNCIL AND THE EUROPEAN ECONOMIC AND SOCIAL COMMITTEE European Union Strategic Approach to Pharmaceuticals in the Environment COM/2019/128.

European Commission (2020a) Commission Implementing Decision (EU) 2020/1161 of 4 August 2020 establishing a watch list of substances for Union-wide monitoring in the field of water policy pursuant to Directive 2008/105/EC of the European Parliament and of the Council (notified under document number C(2020) 5205) (Text with EEA relevance)C/2020/5205.

European Commission (2020b) COMMUNICATION FROM THE COMMISSION TO THE EUROPEAN PARLIAMENT, THE COUNCIL, THE EUROPEAN ECONOMIC AND SOCIAL COMMITTEE AND THE COMMITTEE OF THE REGIONS Pharmaceutical Strategy for Europe COM/2020/761. Brussels.

Fallah, Z., Zare, E. N., Ghomi, M., Ahmadijokani, F., Amini, M., Tajbakhsh, M., Arjmand, M., Sharma, G., Ali, H., Ahmad, A., Makvandi, P., Lichtfouse, E., Sillanpää, M. & Varma, R. S. (2021) Toxicity and remediation of pharmaceuticals and pesticides using metal oxides and carbon nanomaterials. *Chemosphere*, 275, 130055.

Femina Carolin, C., Senthil Kumar, P., Janet Joshiba, G. & Vinoth Kumar, V. (2021) Analysis and removal of pharmaceutical residues from wastewater using membrane bioreactors: A review. *Environmental Chemistry Letters*, 19(1), 329–343.

Fonte, E., Ferreira, P. & Guilhermino, L. (2016) Temperature rise and microplastics interact with the toxicity of the antibiotic cefalexin to juveniles of the common goby (Pomatoschistus microps): Post-exposure predatory behaviour, acetylcholinesterase activity and lipid peroxidation. *Aquat Toxicol*, 180, 173–185.

González Peña, O. I., López Zavala, M. Á. & Cabral Ruelas, H. (2021) Pharmaceuticals market, consumption trends and disease incidence are not driving the pharmaceutical research on water and wastewater. *International journal of environmental research and public health*, 18(5), 2532.

HCWH (2021) *Safer Pharma*, 2021. Available online: https://noharm-europe.org/issues/europe/safer-pharma [Accessed.

Hena, S., Gutierrez, L. & Croué, J.-P. (2021) Removal of pharmaceutical and personal care products (PPCPs) from wastewater using microalgae: A review. *Journal of Hazardous Materials*, 403, 124041.

HPRC (2020) *Solutions for Hospitals*, 2020. Available online: https://www.hprc.org/hospitals [Accessed.

Jenke, D. (2015a) Development and justification of a risk evaluation matrix to guide chemical testing necessary to select and qualify plastic components used in production systems for pharmaceutical products. *PDA Journal of Pharmaceutical Science and Technology*, 69(6), 677.

Jenke, D. (2015b) Moving forward towards standardized analytical methods for extractables and leachables profiling studies. *PDA Journal of Pharmaceutical Science and Technology*, 69(4), 471.

Jenke, D. (2021) A safety risk-based extractables and/or leachables qualification strategy for packaged drug products. *PDA Journal of Pharmaceutical Science and Technology*, 2020, 012617.

Khan, N. A., Vambol, V., Vambol, S., Bolibrukh, B., Sillanpaa, M., Changani, F., Esrafili, A. & Yousefi, M. (2021) Hospital effluent guidelines and legislation scenario around the globe: A critical review. *Journal of Environmental Chemical Engineering*, 9(5), 105874.

Krishnan, R. Y., Manikandan, S., Subbaiya, R., Biruntha, M., Govarthanan, M. & Karmegam, N. (2021) Removal of emerging micropollutants originating from pharmaceuticals and personal care products (PPCPs) in water and wastewater by advanced oxidation processes: A review. *Environmental Technology & Innovation*, 23, 101757.

Maculewicz, J., Kowalska, D., Świacka, K., Toński, M., Stepnowski, P., Białk-Bielińska, A. & Dołżonek, J. (2022) Transformation products of pharmaceuticals in the environment: Their fate, (eco) toxicity and bioaccumulation potential. *Science of The Total Environment*, 802, 149916.

Matamoros, V., Arias, C., Brix, H. & Bayona, J. M. (2009) Preliminary screening of small-scale domestic wastewater treatment systems for removal of pharmaceutical and personal care products. *Water Research*, 43(1), 55–62.

McDonald, G. J. (2019). Single-Use Technology Has Biopharma Sizing Up the Recycling Bin. *Genetic Engineering and Biotechnology News*, 39(8), 50-53. 10.1089/gen.39.08.15

Nguyen, P. Y., Carvalho, G., Reis, M. A. M. & Oehmen, A. (2021) A review of the biotransformations of priority pharmaceuticals in biological wastewater treatment processes. *Water Research*, 188, 116446.

Origin Pharma Packaging (2021) *Trends in Pharmaceutical Packaging 2021*, 2021. Available online: https://www.pharmaceutical-technology.com/contractors/packaging/origin/pressreleases/pharmaceutical-packaging-2021 [Accessed.

O'Neill, S. M. & Lawler, J. (2021) Knowledge gaps on micro and nanoplastics and human health: A critical review. *Case Studies in Chemical and Environmental Engineering*, 3, 100091.

Patel, M., Kumar, R., Kishor, K., Mlsna, T., Pittman, C. U. & Mohan, D. (2019) Pharmaceuticals of emerging concern in aquatic systems: Chemistry, occurrence, effects, and removal methods. *Chemical Reviews*, 119(6), 3510–3673.

Prata, J. C. (2018) Microplastics in wastewater: State of the knowledge on sources, fate and solutions. *Marine Pollution Bulletin*, 129(1), 262–265.

Prata, J. C., Lavorante, B., MDC, B. S. M. M. & Guilhermino, L. (2018) Influence of microplastics on the toxicity of the pharmaceuticals procainamide and doxycycline on the marine microalgae Tetraselmis chuii. *Aquat Toxicol*, 197, 143–152.

Puckowski, A., Cwięk, W., Mioduszewska, K., Stepnowski, P. & Białk-Bielińska, A. (2021) Sorption of pharmaceuticals on the surface of microplastics. *Chemosphere*, 263, 127976.

Qu, H., Ma, R., Wang, B., Yang, J., Duan, L. & Yu, G. (2019) Enantiospecific toxicity, distribution and bioaccumulation of chiral antidepressant venlafaxine and its metabolite in loach (*Misgurnus anguillicaudatus*) co-exposed to microplastic and the drugs. *J Hazard Mater*, 370, 203–211.

Research and Markets (2021) *Plastic Packaging Market - Growth, Trends, COVID-19 Impact, and Forecasts (2021 - 2026)*, 2021. Available online: https://www.researchandmarkets.com/reports/4771760/plastic-packaging-market-growth-trends-covid [Accessed.

Santana-Viera, S., Montesdeoca-Esponda, S., Torres-Padrón, M. E., Sosa-Ferrera, Z. & Santana-Rodríguez, J. J. (2021) An assessment of the concentration of pharmaceuticals adsorbed on microplastics. *Chemosphere*, 266, 129007.

Shruti, V. C., Pérez-Guevara, F., Elizalde-Martínez, I. & Kutralam-Muniasamy, G. (2020) First study of its kind on the microplastic contamination of soft drinks, cold tea and energy drinks - Future research and environmental considerations. *Science of The Total Environment*, 726, 138580.

Stock, V., Böhmert, L., Lisicki, E., Block, R., Cara-Carmona, J., Pack, L. K., Selb, R., Lichtenstein, D., Voss, L., Henderson, C. J., Zabinsky, E., Sieg, H., Braeuning, A. & Lampen, A. (2019) Uptake and effects of orally ingested polystyrene microplastic particles in vitro and in vivo. *Archives of Toxicology*, 93(7), 1817–1833.

Stock, V., Fahrenson, C., Thuenemann, A., Dönmez, M. H., Voss, L., Böhmert, L., Braeuning, A., Lampen, A. & Sieg, H. (2020) Impact of artificial digestion on the sizes and shapes of microplastic particles. *Food and Chemical Toxicology*, 135, 111010.

Stock, V., Laurisch, C., Franke, J., Donmez, M. H., Voss, L., Bohmert, L., Braeuning, A. & Sieg, H. (2021) Uptake and cellular effects of PE, PP, PET and PVC microplastic particles. *Toxicology in Vitro*, 70.

Transparency Market Research (2021) *Pharmaceutical Plastic Packaging Market- Global Industry Analysis, Size, Share, Growth, Trends and Forecast 2017 - 2025*, 2021. Available online: https://www.transp arencymarketresearch.com/pharmaceutical-plastic-packaging-market.html [Accessed.

Verlicchi, P., Galletti, A., Petrovic, M., Barceló, D., Al Aukidy, M. & Zambello, E. (2013) Removal of selected pharmaceuticals from domestic wastewater in an activated sludge system followed by a horizontal subsurface flow bed — Analysis of their respective contributions. *Science of The Total Environment*, 454-455, 411–425.

Vieira, Y., Lima, E. C., Foletto, E. L. & Dotto, G. L. (2021) Microplastics physicochemical properties, specific adsorption modeling and their interaction with pharmaceuticals and other emerging contaminants. *Sci Total Environ*, 753, 141981.

World Health Organisation (2021) *GHE: Life expectancy and healthy life expectancy*, 2021. Available online: https://www.who.int/data/ gho/data/themes/mortality-and-global-health-estimates/ghe-life-expectancy-and-healthy-life-expectancy [Accessed.

Wu, B., Wu, X., Liu, S., Wang, Z. & Chen, L. (2019) Size-dependent effects of polystyrene microplastics on cytotoxicity and efflux pump inhibition in human Caco-2 cells. *Chemosphere*, 221, 333–341.

Wu, C., Zhang, K., Huang, X. & Liu, J. (2016) Sorption of pharmaceuticals and personal care products to polyethylene debris. *Environ Sci Pollut Res Int*, 23(9), 8819–8826.

Xu, B., Huang, D., Liu, F., Alfaro, D., Lu, Z., Tang, C., Gan, J. & Xu, J. (2021) Contrasting effects of microplastics on sorption of diazepam and phenanthrene in soil. *Journal of Hazardous Materials*, 406, 124312.

Zaied, B. K., Rashid, M., Nasrullah, M., Zularisam, A. W., Pant, D. & Singh, L. (2020) A comprehensive review on contaminants removal from pharmaceutical wastewater by electrocoagulation process. *Science of the Total Environment*, 726, 138095.

Zhang, S., Ding, J., Razanajatovo, R. M., Jiang, H., Zou, H. & Zhu, W. (2019) Interactive effects of polystyrene microplastics and roxithromycin on bioaccumulation and biochemical status in the freshwater fish red tilapia (*Oreochromis niloticus*). *Sci Total Environ*, 648, 1431–1439.

Zhou, R., Lu, G., Yan, Z., Jiang, R., Bao, X. & Lu, P. (2020) A review of the influences of microplastics on toxicity and transgenerational effects of pharmaceutical and personal care products in aquatic environment. *Science of the Total Environ*, 732, 139222.

Zuccarello, P., Ferrante, M., Cristaldi, A., Copat, C., Grasso, A., Sangregorio, D., Fiore, M. & Conti, G. O. (2019) Exposure to microplastics (<10 mu m) associated to plastic bottles mineral water consumption: The first quantitative study. *Water Research*, 157, 365–371.

CHAPTER 5

Fungal Bioremediation of Microplastics

Naveen Kumar[1], Suresh C. Pillai[2], and Mary Heneghan[1,3]

[1]*Fungal Molecular Biology Group, Institute of Technology Sligo, Ash Lane, SligoF91 YW50, Ireland*
[2]*Nanotechnology & Bioengineering Research Group (Nano-Bio), Institute of Technology Sligo, Ash Lane, SligoF91 YW50, Ireland*
[3]*Centre for Environmental Research, Innovation and Sustainability (CERIS), Institute of Technology Sligo, Ash Lane, SligoF91 YW50, Ireland*

CONTENTS

DOI: 10.1201/9781003109730-5

5.1 INTRODUCTION

Plastics have become an inevitable part of human life since their discovery in the early 1900s. This is due to their versatile nature, cost, and extensive applications in various industries. Global production of plastics in 2019 was 368 million tonnes, with 16% of this produced in Europe (Plastics Europe, 2020). While plastics have played an important and beneficial role in revolutionizing certain industries, they have also become a major concern for environmental health. Plastics are now considered a ubiquitous pollutant, found in the remotest locations on Earth, such as, the Marina Trench and in Arctic ice cores (Bergmann et al, 2017; Huntington et al, 2020).

Microplastics (MPs) have received increasing attention within both the mainstream media and the scientific community since they were first described by Thompson et al. (2004). MPs are plastics sized <5 mm, including nanosized plastics <1 μm (NPs), and are broadly categorized into two groups based on their formation: primary and secondary. Primary MPs are purposely synthesized for specific applications, such as exfoliants in personal care products (Kane & Clare, 2019). Secondary MPs result from the breakdown of larger plastic items by physical, chemical, and biological mechanisms. Furthermore, MPs vary in both size and shape and can be described as fibres, spheres, and/or fragments (Enfrin et al, 2021).

MPs present in the environment are composed of chemical mixtures as opposed to pure polymers (Thompson et al, 2009). It has been suggested that MPs may leach chemicals that were incorporated during their synthesis, as they are not bound within the polymer matrix, and present a threat to aquatic life (Barnes et al, 2009). The complexity in composition may also lead to additional detrimental effects to the environment through other mechanisms as MPs may absorb and concentrate hydrophobic organic contaminants (Lee et al, 2014; Velzeboer et al, 2014) or accumulate other pollutants such as heavy metals (Gallo et al, 2018).

It is widely acknowledged that humans are exposed to MPs from daily use commodities, food, air, and water (Smith et al, 2018). They have been reported in processed foods, beverages, and sea foods (Wright & Kelly, 2017). They have been detected in freshwater, marine water, wastewater, and drinking water (both bottled and tap water) (World Health Organization, 2019). MPs have been found in sludge by-products which are used in farming applications, leading to contamination of soil, water, and food (Wang et al, 2019). It was reported that in the United States, 8 trillion microbeads enter water bodies every day due to ineffective wastewater treatment plants (Rochman et al, 2015). More recent studies have suggested that this figure may be less, but could still result in approximately 65 million MPs being released on a daily basis (Hamidian et al, 2021). MPs have been ingested by a range of organisms once they enter aquatic systems. This enables MPs to accumulate throughout the food web, which ultimately can cause harm to humans and the environment (Environmental Protection Agency (EPA), 2017).

While daily exposure to MPs is evident, the impact and level of risk to human health are still unknown and are under intense investigation (Katyal et al, 2020). A key area that needs to be addressed in order to answer the question of human exposure and risk is the development of adequate tools for sampling, isolation, quantification, and characterization of MPs (Blair et al, 2019; Velimirovic et al, 2021).

The growing environmental problems associated with plastic production have led to the manufacture of biodegradable plastics as a potential solution to alleviate the accumulation of plastic waste in the ecosystem (Haider et al, 2018). Biodegradable plastics are polymeric materials made from either renewable raw materials or from petrochemicals. They can be decomposed into carbon dioxide, methane, water, inorganic compounds, or biomass either by hydrolysis, ultraviolet (UV) light degradation, or the action of microorganisms (GESAMP, 2016). The rate at which they degrade is highly variable (Chinaglia et al, 2018), with multiple factors influencing this breakdown. These factors include chemical structure, molecular weight, and surface area of the polymer (Al Hosni et al, 2019). Crucially, the degradation rate will depend on the fate of the biodegradable plastic, i.e., the environment that it ends up in, such as soil or marine water. Some biodegradable plastics require an extended period of time at temperatures of 50°C or above in order to degrade completely. While these conditions exist in industrial composting units, they are rarely met in the natural environment. Relatively little is known about the time required for biodegradable plastics to fully degrade in soil or compost, or the biotic and abiotic conditions that facilitate efficient degradation. Given the current scale of the pollution problem and the complex nature of the polymers used, techniques for achieving consistent degradation are required. The use of biodegradable plastics as an alternative can only be achieved if proper waste management channels, garbage disposal, and industrial biodegradation facilities are set up at a considerable scale.

Biological degradation of plastics and biodegradable plastics has received significant attention in the last decade as the scientific community strives to reverse the global trend of rising plastic waste and MP accumulation in the environment. Bioremediation is the use of microorganisms, enzymes, or plants (Glazer & Nikaido, 1995) to breakdown pollutants or decrease the associated risks for humans and the environment (Baranger et al, 2021; Gaur, Narasimhulu & Y, 2018). It is increasingly used to alleviate environmental accidents and is mainly applied to soil, sludges, and several types of residual waters (Zhang et al, 2017; Jenkins et al, 2019; Quintella, Mata & Lima, 2019; Hwang et al, 2020; da Silva & Gouveia, 2020). It is regarded as an environmentally friendly and cost-effective option which can be applied *in situ* (Das and Adholeya, 2021; Oliveira et al, 2020).

Numerous studies have investigated microbial degradation of plastics but have mainly focused on bacteria (Urbanek et al, 2018, Rana, 2019; Yuan et al, 2020). Fungi are very promising organisms for bioremediation of MPs and biodegradable plastics due to their ubiquity in nature, the array of powerful enzymes they possess, and their ability to adapt to the microplastic-microorganism ecosystem (Kettner et al, 2017). Recent studies have shown fungi to be an efficient and rapid degrader of both plastics (Spina et al, 2021; Quintella, Mata & Lima, 2019) and biodegradable plastics (Muhonja et al, 2018; Sankhla et al, 2020). This chapter will evaluate the interdisciplinary literature available to understand and assess the potential of fungi in MP degradation and will draw recommendations for future research and environmental policies.

5.2 BIOTECHNOLOGICAL AND ECOLOGICAL SIGNIFICANCE OF FUNGI

Fungi have existed on earth for at least one thousand million years (Heckman et al. 2001). The kingdom Fungi (superkingdom eukaryotes) is predicted to comprise millions of species (Badotti et al, 2018) which includes mushrooms, yeasts, moulds, smuts, rusts, and puffballs. Fungi are distinctly different from the animal and plant kingdom due to the presence of chitin in their cell wall and their mode of nutrient acquisition, which is absorptive rather than digestive (Walker & White, 2017). These organisms have a clear ecological role, where the depolymerization of biopolymers is a key process in the cycling of carbon, with litter decomposition in temperate forests mainly driven by fungal activity. They constitute a major fraction of the living biomass responsible for efficient degradation of many recalcitrant organic compounds in soil litter and the humic layer (Bhatnagar et al, 2018). Fungi are economically significant, not only in terms of mushroom production cultivated for human consumption but they are also employed in many industries such as brewing and baking. They play an important role in many biotechnological processes, including the production of antibiotics, alcohols, enzymes, organic acids, and numerous pharmaceuticals (Zhang et al, 2018). The discovery of recombinant DNA technology and large-scale genomic analysis has increased the use of yeasts and filamentous fungi in commercial applications (Chan et al, 2018).

5.3 FUNGI AS CANDIDATE ORGANISMS FOR BIOREMEDIATION OF MICROPLASTICS

Fungi are a diverse and abundant group of organisms belonging to the kingdom Eumycota, representing one of the five main multicellular lineages in the tree of life (Viegas et al, 2016). Fungal classification is in a constant state of flux and numerous different classifications have been proposed (Hawksworth, 1991, Hibbett et al, 2007). The most recent reclassification by Tedersoo et al. (2018) utilized a comprehensive system of phylogenetic, monophyletic, and divergence time approaches and recognizes a kingdom (Fungi), nine subkingdoms; 18 phyla and 23 subphyla (Tedersoo et al, 2018). Similar to plants and animals, these organisms possess significant attributes of complex multicellularity such as cell–cell communication, cell–cell adhesion, long-range transport, programmed cell death, and a developmental lifecycle (Boussard et al, 2019).

Fungi are heterotrophic organisms and feed by utilizing nutrients outside their cell. They can obtain their food from dead organic matter (saprophytes), by killing other organisms (parasites), by engaging in symbiotic/mutualistic relationships with plants, insects, or animals or by a combination of these methods (Sánchez, 2020). Consequently, fungal growth relies on the secretion of extracellular proteins involved in sensing, signalling, nutrient acquisition, and cell wall building (Filiatrault-Chastel et al, 2021). Fungi are recognized as the microorganisms responsible for the degradation of most organic compounds in the environment. They release digestive enzymes by exocytosis that breakdown macro and organic molecules into simpler organic compounds. Absorption of the compounds is accompanied by release of CO_2 and H_2O under aerobic conditions, or CH_4 under anaerobic conditions (Pathak & Navneet, 2017).

Fungi have an ability to adapt and grow in harsh environments (Newsham, 2012). In particular, filamentous fungi have been shown to respond to variable environmental conditions and exhibit tolerance towards an array of pollutants (Olicón-Hernández et al, 2017). These fungi have the ability to modify their secretome (entire set of proteins secreted by an organism under given circumstances) and thus their metabolism in response to changing environmental conditions (Filiatrault-Chastel et al, 2021).

Fungi use their filamentous structure for exploring and growing in difficult places giving them a competitive edge over other microorganisms. This hyphal (filamentous) structure enables them to combine biochemical and physical actions, using secondary metabolites as enzymes and biosurfactants (Sánchez, 2020; Kim & Rhee, 2003). Furthermore, the hyphal structure promotes water purification through biodsorption, where contaminants adhere to their surface or can be internalized in the cell, and thus removed from water systems (Kumar & Min, 2011; Lu et al, 2016).

Fungi are excellent candidates for bioremediation due to their powerful enzymatic system, production of biosurfactants (i.e., hydrophobins), and their adsorption capabilities. They can synthesize an array of intracellular and extracellular enzymes that can degrade a variety of complex polymers, including plastics. Fungi exhibit a high ability to grow on diverse surfaces, forming biofilms (enabled by their absorptive nutrition mode, secretion of extracellular enzymes that digest complex molecules, and apical hyphal growth) (Afonso et al, 2021). Recent studies have shown fungi have the capacity to adhere to and utilize MPs as a carbon source for degradation (Woodall et al, 2014; Yuan et al, 2020;)

5.3.1 Fungal Enzyme Systems

Fungi adapt to extreme conditions colonizing different matrices and degrading organic waste matter. To facilitate this, these organisms have an arsenal of both intracellular and extracellular enzyme pathways that can degrade recaliant compounds. Saprotrophic fungi in particular produce a range of extracellular enzymes for the degradation of complex plant polymers including lignin, cellulose, and hemicellulose. Due to the low specificity of this enzymatic machinery, other targets, including microplastics and biodegradable plastics, can be broken down (Akerman-Sanchez & Rojas-Jimenez, 2021). The battery of enzymes released in the fungal secretome includes glycoside hydrolases, polysaccharide lyases, carbohydrate esterases, oxidoreductases, and hydrogen peroxide-producing enzymes (Bouws, Wattenberg & Zorn, 2008).

Fungal enzyme classes that are specifically associated with plastic degradation include laccases, class II peroxidases, esterases, proteases, lipases, and cutinases (Sánchez, 2020). While enzyme activities involved in plastic degradation are mainly hydrolytic, oxidoreductases also play a key role. These enzymes can oxidize a variety of substrates by transferring electrons from organic compounds to molecular oxygen (Daly et al, 2021).

Intracellular enzymes play a crucial role in the internal mechanism of detoxification and facilitate adaptation to unfavourable conditions (Schwartz et al, 2018). Fungal Cytochrome P450 mono-oxygenases (CYP 450) are a superfamily of heme-thiolate enzymes that are involved in a range of biochemical pathways. They have been identified in hydroxylation, sulfoxidation, desulfuration, dehalogenation, deamination, and epoxidation reactions which are critical for cell function (Shin et al, 2018). Their activity requires a short electron transport chain shuttle of electrons from either NADH or NADPH to the oxygenase (McLean et al., 2005). These enzymes help in primary metabolism, hyphal wall integrity, and formation of outer spore wall (Črešnar & Petrič, 2011).

Fungi have a diverse family of CYP separate from bacteria, animals, or plants (Guan et al, 2017). CYP 450 enzymes catalyze the conversion of hydrophobic intermediates of metabolic pathways (primary and secondary) and enable the growth of fungi under different conditions; by detoxifying the environmental pollutants (Chen et al, 2014). They play a key role in phase I and phase II of xenobiotic metabolism. Specifically, epoxidases and transferases have been identified in phase I and phase II metabolism respectively. Epoxidases catalyze oxidation reactions in phase I metabolism while transferases are involved in conjugation reactions in phase II metabolism (Hernández-Arenas et al, 2021).

5.3.2 Fungal Hydrophobins

Fungal hydrophobins play a critical role in the degradation processes, as they act as biosurfactants, increasing bioavailability, and surface mobility (Sánchez, 2020). They are amphipathic proteins made up of eight conserved cysteine residues linked by disulfide bonds (Wösten & Scholtmeijer, 2015). These proteins are composed of 70–350 amino acids and are responsible for the formation of aerial structures, such as spores, hyphae, and fruiting bodies, in filamentous fungi. Crucially, they are involved in the attachment of hyphae to hydrophobic surfaces (Wessels, 1997).

Hydrophobins pre-fabricate monolayers on interaction with hydrophobic and hydrophilic surfaces. They are classified based on the morphology of their monolayer, hydrophobic pattern, and solubility in detergents. Class I aqueously insoluble hydrophobins form functional amyloid fibres with rodlet morphology while Class II aqueous soluble hydrophobins dissolve in organic solvents and create monolayers absent of rodlet morphology (Wu et al, 2017).

The unique properties of fungal hydrophobins such as their strong adhesion, surface activity, and formation of highly insoluble self-assembled structures have attracted significant attention not only for bioremediation but for applications in food science (Cox & Hooley, 2009), drug delivery (Fang et al, 2014), biosensor development (Della Ventura et al, 2016; Tao et al, 2017), and coatings for medical implants (Olicón-Hernández et al, 2017). Furthermore, hydrophobin coatings have been reported to improve fibroblast growth and morphology on plastic surfaces such as Teflon and polystyrene (Janssen et al, 2004; Piscitelli et al, 2017).

5.4 FACTORS AFFECTING MICROBIAL DEGRADATION OF PLASTICS/MICROPLASTICS

Fungal degradation of plastics/MPs is dependent on three main factors: microbial properties, plastic/MP characteristics, and environmental conditions. Microbial properties include the type of fungus (basidiomycete, ascomycete, filamentous, yeast, etc.), growth conditions (temperature, pH, oxygen concentration, nutrients, etc.), and types of enzymes produced.

Plastic/MP characteristics include physical and chemical properties such as surface area, hydrophilic/hydrophobic properties, molecular weight, density, substituents, functional groups, melting temperature, and crystallinity of the compound (Mohanan et al, 2020). For example, the action of enzymes on plastic is influenced by the density of plastic. Balasubramanian et al. (2010) reported that high-density polyethylene (HDPE) was degraded significantly more than low-density polyethylene (LDPE) when treated with the same group of enzymes.

Environmental factors such as UV, humidity, pH, heat, and chemical factors play a vital role in biodegradation (Khan et al, 2017). Firstly, environmental conditions act as a catalyst for the growth and metabolism of fungi. A shift in temperature or pH can affect the growth of the fungus. Secondly, these conditions will have an impact on plastic waste in the ecosystem, making them more susceptible to microbial degradation.

An oxidizing environment causes aging of MPs which increases the efficiency of fungal enzymes in degrading these MPs. UV irradiation has been shown to play a significant role in MP degradation by decreasing the molecular weight and tensile properties of the polymer (Jeon et al, 2016). UV irradiation produces photo-oxidants which synergize with the enzymes degradation activity and hence it's easier for the microbe to degrade the MP (Falkenstein et al, 2020). MPs possessing a chromophore group are more susceptible to microbial degradation when UV irradiation is used as a co-treatment (Yamada-Onodera et al, 2001).

Chemical factors, such as the presence of ethanol, have been reported to benefit LDPE plastic degradation. Additionally, additives and plasticizers (organic acid esters) are commonly used in polyvinyl chloride (PVC) plastics, such as di-isononyl phthalate (DNP) and di-ethylhexyl phthalate (DEHP), for modification of the physical and mechanical properties of the polymer. Fungi can degrade ester-based plasticizers using lipases and hydrolases (Danso et al, 2019). Additives, such as antioxidants, heat, and light stabilizers, also have an impact on biodegradation and can increase the rate of degradation (Yousif & Haddad, 2013).

In summary, there are numerous factors that influence MP degradation, and it is imperative that the optimal physical and chemical conditions for MP degradation under specific environmental conditions are explored and understood.

5.5 FUNGAL DEGRADATION OF PLASTICS AND MICROPLASTICS

The potential for microorganisms, such as bacteria and fungi, to degrade plastics, is well documented (Rana, 2019; Yuan et al, 2020). Many review papers are available that describe in detail the different factors affecting fungal degradation of plastics along with the hydrolytic and oxidative enzyme systems involved

(Afonso, Simões & Lima, 2021; Akerman-Sanchez & Rojas-Jimenez, 2021; Daly et al, 2021; Sánchez, 2020). Microbial degradation of diverse types of plastics such as polypropylene (PP), polyethylene (PE), polyethylene terephthalate (PET), polystyrene (PS), and polyvinyl chloride (PVC), have been reported (Ameen et al, 2015; Danso et al, 2019; Muenmee et al, 2016; Auta et al, 2018; Giacomucci et al, 2019). Furthermore, studies have also shown fungi to be an efficient and rapid degrader of biodegradable plastics (Muhonja et al, 2018; Sankhla et al, 2020). This section provides a brief overview of the different plastics and MPs that have been degraded by fungi.

Plastic biodegradation is measured by evaluation of microbial growth and/or by monitoring the changes in the polymer properties. Fungal growth in aerobic conditions is assessed by biomass production, biochemical oxygen demand (BOD), enzyme assays, and CO_2 production. Fungal enzymes effect biodegradation by altering the crystallinity, molecular weight, degree of fragmentation, tensile properties, and the functional groups present on the plastic surface. Research has shown that the presence of hydrolyzable ester bonds in the plastic increases the biodegradability of the polymer (Rydz et al, 2014).

Polyurethane (PUR) is a versatile class of synthetic polymer materials used in many industries, such as medical devices, buildings, and construction, etc. PUR can be synthesized (by reacting isocyanates and polyols) with tailored physical properties providing enormous flexibility in its applications. It is the 6th most abundant plastic worldwide (Liu et al, 2021). Numerous fungal species have been reported to degrade PUR, including strains from the genera *Aspergillus, Pestalotiopsis, Cladosporium, Fusarium*, and *Penicillium* (Magnin et al, 2019; Magnin et al, 2020). Strains from the species *Monascus* have been shown to biodegrade PUR by producing proteases, lipases, and esterases (Loredo-Treviño et al, 2012). *Pestalotiopsis* microspores showed enhanced serine hydrolase activity when the fungus was grown with impranil as the carbon source (Russell et al, 2011).

Polyethylene (PE) is the most common form of plastic used for packaging. PE is formed by a mixture of analogous polymers of ethylene. It is classified in two groups, high-density polyethylene (HDPE) and low-density polyethylene (LDPE), depending upon the number of ethylene polymers. Its vast production and widespread usage is creating pollution and has attracted the interest of researchers. Ameen et al, (2015) reported that a consortium of ascomycete strains are able to degrade LDPE by producing laccases, lignin peroxidases (LiP), and manganese peroxidases (MnP). These enzymes cleave the length of the PE chain, which offers insight into the metabolic pathway for the degradation of PE.

Polypropylene (PP) is used in cosmetics and personal care products. Several microorganisms have been identified as efficient degrader of this polymer. Microbial enzymes such as depolymerase, esterase, and ligninolytic enzymes can cleave these polymers into monomers; however, there is no evidence available in the literature regarding the mechanism of this degradation. White rot fungi (WRF), including *P. ostreatus* have been studied for biodegradation of PP. *Aspergillus terreus* and *Aspergillus sydowii* strain have also been identified as efficient PP degraders (Auta et al, 2018; Hamidian et al, 2021).

PET plastics are a major source of plastic pollution due to their wide presence in the form of debris from plastic bottles. PET degradation is usually done by (a) abiotic degradation processes like hydrolysis, thermal degradation, and chemical degradation and (b) biotic/ biodegradation. Biodegradation of PET typically occurs via serine hydrolases such as cutinases, lipases, and carboxylesterases (Danso et al, 2019; Hamidian et al, 2021). Two enzymes mono hydroxyethyl terephthalate hydrolase (MHETase) and polyethylene terephthalate hydrolase (PETase) showed efficient PET degradation. These enzymes work in sequence, where PETase degrades PET into monomeric mono-2-hydroxyethyl terephthalate (MHET) and MHETase then converts MHET into terephthalic acid (TPA) and ethylene glycol

(Palm et al, 2019). Studies by Kawai et al. (2019) also reported the potential of polyester hydrolase and cutinase to degrade PET. Strains of *Penicillium* and *Thermomyces* demonstrated an ability to hydrolyze PET pellets, while cutinase activity was detected in *Fusarium solani* when 7% PET film was used as a substrate (Kawai et al, 2019).

Polystyrene (PS) is primarily used in the food packaging industry in two forms, expanded polystyrene (EPS) or extruded polystyrene (XPS). Various organisms have been reported to degrade polystyrene. Enzymes employed in PS degradation include styrene oxidase isomerase, styrene monooxygenase, and phenyl acetyl coenzyme. They act by reducing PS to its monomer styrene, which then undergoes oxidation before entry into the TCA cycle. Fungal species reported as efficient degraders of PS include *Cephalosporium sp*, *Mucor sp* (Muenmee et al, 2016), and *Lentinus tigrinus* (Tahir et al, 2013).

PVC is a synthetic thermoplastic that is widely used for its unique plasticized form. Few studies have been performed on fungal degradation of PVC and to date, only the WRF have shown activity against this plastic (Ali et al, 2014).

Limited studies have been described for fungal-mediated bioremediation of MPs (Table 5.1), highlighting the novel research opportunity to find efficient MP degrading strains. Fungi have been found to alter the morphology as well as internal properties of MPs (Yuan et al, 2020). Fungal classes ascomycetes, basidiomycetes, and zygomycetes have been reported for their excellent degradation capacity of MPs present in petroleum. The marine ascomycetes *Gloeophyllum trabeum*, *Zalerion maritimum*, are able to utilize polyethylene by producing laccase and manganese independent peroxidase enzymes, with minimum external nutrients (Krueger et al, 2015). Russell et al, (2011) demonstrated that the fungal enzyme serine hydrolase can degrade MPs.

The biodegradation pathways employed by fungi for MPs are not well defined (Sangeetha Devi et al, 2015). Hypothetical PE and PET MP degradation pathways have been proposed based

TABLE 5.1 Degradation of microplastics by fungi

Microbial taxa	MP type	Duration of degradation	Findings	References
Penicillium simplicissimum YK	PE irradiated with UV light	3 months	Polyethylene molecular weight reduced	(Yamada-Onodera et al, 2001)
Penicillium pinophilum ATCC 11797	LDPE powder	3 months	Morphological and structural changes observed	(Volke-Seplveda et al, 2002)
Zalerion maritimum	PE pellets	28 days	Weight loss and structural changes observed	(Paço et al, 2017)
Pleurotus ostreatus	PE	2 months	No significant polymer reduction	(Da Luz et al, 2015)
Pestalotiopsis microspore	PS	1 month	Weight loss observed	(Russell et al, 2011)
Aspergillus oryzae	PE	4 months	Weight loss detected	(Muhonja et al, 2018)
Aspergillus flavus VRKPT1	HDPE	1 month	Weight loss observed	(Sangeetha Devi et al, 2015)
Thermomyces lanuginosus	PLA	2 months	No significant polymer reduction	(Karamanlioglu et al, 2014)
Phanerochaete chrysosporium	PVC	6 months	No significant polymer reduction	(Fakhrul et al, 2014)
Trichoderma harzianum	PE	3 months	Weight loss shown, suggests 40% polymer reduction	(Sowmya et al, 2014)

on fungal enzyme chemistry, their reaction with substrates, and biosurfactant production ability (Das & Chandran, 2011; Sánchez, 2020). The putative pathway described for *Aspergillus terreus* under aerobic conditions is based on the catalytic activity of heme manganese peroxidase (heme MnP) (Balasubramanian et al., 2014). Conversely, the proposed pathway for PET degradation by *Fusarium solani* involves the activation of cutinases (Danso et al, 2019; Kettner et al, 2017).

5.6 CONCLUSION

Plastic waste has been a long-standing cause for concern with respect to the diversified presence and usage of plastic globally. The widespread use and subsequent breakdown of plastics lead to the formation of micro and nanoplastics. These small-sized plastics are widely affecting humans, animals, and the environment due to their ubiquitous presence and inconspicuous size. The consequences of this to human health is the subject of ongoing research. There is an urgent need for action, and the plastic waste problem has become a priority for the G7 leaders who have acknowledged it as a "global challenge, directly affecting marine, and coastal life and ecosystems and potentially human health" (G7 Summit, 2015). A lifecycle approach to include design, production, consumption, end-of-life management, and disposal of plastics has been suggested as a strategy to significantly reduce the amount of plastic waste (G7 Environment Ministers' Meeting, 2019).

Bioremediation offers an eco-friendly and economically viable option for waste management of plastics. Fungi are well-equipped organisms for the degradation of recalcitrant molecules, with their unique enzymatic systems, hydrophobin substrate-binding mechanisms, non-specificity of substrates, and their capacity to penetrate three-dimensional structures. Fungi have proved their efficacy in degrading different types of plastics such as PET, PS, and PE; however, limited knowledge is available regarding the metabolic pathways and enzymes involved in the degradation process. Furthermore, degradation of MPs like PP and PVC is not well explored. More studies are needed to understand the enzyme reactions involved in their degradation and to isolate, purify, and characterize the enzymes.

Fungi represent an underutilized resource for the degradation of MPs. These mighty organisms offer a green and sustainable option for tackling plastic waste pollution. Future research is needed in this area to unlock the full biochemical potential of fungi for biotransformation of plastics. Successful mycoremediation will also rely heavily on selection of the correct fungus to target a particular MP. Advanced molecular engineering, fungal databases, and proteomic approaches will help clarify and solve the obstacles in the biodegradation of complex polymers.

REFERENCES

Al Hosni, A., Pittman, J. & Robson, G. (2019). Microbial degradation of four biodegradable polymers in soil and compost demonstrating polycaprolactone as an ideal compostable plastic. *Waste Management*, 97, 105–114.

Afonso, T., Simões, L. & Lima, N. (2021). Occurrence of filamentous fungi in drinking water: Their role on fungal-bacterial biofilm formation. *Research in Microbiology*, 172(1).., p.103791.

Akerman-Sanchez, G. & Rojas-Jimenez, K. (2021). Fungi for the bioremediation of pharmaceutical-derived pollutants: A bioengineering approach to water treatment. *Environmental Advances*, 4, 100071.

Ali, M.I., Ahmed, S., Javed, I., Ali, N., Atiq, N., Hameed, A. & Robson, G. (2014). Biodegradation of starch blended polyvinyl chloride films by isolated *Phanerochaete chrysosporium* PV1. *Int. J. Environ. Sci. Technol.* 11, 339–348. 10.1007/s13762-013-0220-5.

Ameen, F., Moslem, M., Hadi, S. & Al-Sabri, A. E. (2015). Biodegradation of low density polyethylene (LDPE) by mangrove fungi from the red sea coast. *Progress in Rubber, Plastics and Recycling Technology*, 31(2)... 10.1177/147776061503100204

Auta, H. S., Emenike, C. U., Jayanthi, B. & Fauziah, S. H. (2018). Growth kinetics and biodeterioration of polypropylene microplastics by Bacillus sp. and Rhodococcus sp. isolated from mangrove sediment. *Marine Pollution Bulletin*, 127. 10.1016/j.marpolbul.2017.11.036

Balasubramanian, V., Natarajan, K., Hemambika, B., Ramesh, N., Sumathi, C. S., Kottaimuthu, R. & Rajesh Kannan, V. (2010). High-density polyethylene (HDPE)-degrading potential bacteria from marine ecosystem of Gulf of Mannar, India. *Letters in Applied Microbiology*, 51(2). 10.1111/j.1472-765X.2010.02883.x

Balasubramanian, V. , Natarajan, K. , Rajeshkannan, V. , Perumal, P. (2014). Enhancement of in vitro high-density polyethylene (HDPE) degradation by physical, chemical, and biological treatments. *Environmental Science and Pollution Research*, 21, 12549–12562. 10.1007/s11356-014-3191-2

Baranger, C., Pezron, I., Lins, L., Deleu, M., Le Goff, A. & Fayeulle, A. (2021). A compartmentalized microsystem helps understanding the uptake of benzo[a]pyrene by fungi during soil bioremediation processes. *Science of The Total Environment*, 784, 147151.

Badotti, F., Fonseca, P. L. C., Tomé, L. M. R., Nunes, D. T. & Góes-Neto, A. (2018). 'ITS and secondary biomarkers in fungi: review on the evolution of their use based on scientific publications'. *Brazilian Journal of Botany*, 41 (2)., 471–479.

Barnes, D. K. A., Galgani, F., Thompson, R. C. & Barlaz, M. (2009). Accumulation and fragmentation of plastic debris in global environments. *Philosophical Transactions of the Royal Society B: Biological Sciences*, 364(1526).

Bergmann, M., Wirzberger, V., Krumpen, T., Lorenz, C., Primpke, S., Tekman, M. B. & Gerdts, G. (2017). High quantities of microplastic in arctic deep-sea sediments from the HAUSGARTEN observatory. *Environmental Science and Technology*, 51(19)., 10.1 021/acs.est.7b03331

Bhatnagar, J. M., Peay, K. G. & Treseder, K. K. (2018). 'Litter chemistry influences decomposition through activity of specific microbial functional guilds', *Ecological Monographs*, 88(3), 429-444.

Blair, R. M., Waldron, S., Phoenix, V. R. & Gauchotte-Lindsay, C. (2019). Microscopy and elemental analysis characterisation of microplastics in sediment of a freshwater urban river in Scotland, UK. *Environmental Science and Pollution Research*. 10.1007/s1135 6-019-04678-1

Boussard, A., Delescluse, J., Pérez-Escudero, A. & Dussutour, A. (2019). Memory inception and preservation in slime moulds: the quest for a common mechanism. *Philosophical Transactions of the Royal Society B: Biological Sciences*, 374(1774)., 20180368.

Bouws, H., Wattenberg, A. & Zorn, H. (2008). Fungal secretomes—nature's toolbox for white biotechnology. *Applied Microbiology and Biotechnology*, 80(3).

Chan, L. G., Cohen, J. L. & De Moura Bell, J. M. L. N. (2018). 'Conversion of agricultural streams and food-processing by-products to value-added compounds using filamentous fungi', *Annual review of food science and technology*, 9, 503–523.

Chen, W., Lee, M., Jefcoate, C., Kim, S., Chen, F. & Yu, J. (2014). Fungal cytochrome P450 monooxygenases: their distribution, structure, functions, family expansion, and evolutionary origin. *Genome Biology and Evolution*, 6(7)., 1620–1634.

Chinaglia, S., Tosin, M. & Degli-Innocenti, F. (2018). 'Biodegradation rate of biodegradable plastics at molecular level', *Polymer Degradation and Stability*, 147, 237–244.

Cox, P. W. & Hooley, P. (2009). Hydrophobins: New prospects for biotechnology. *Fungal Biology Reviews*, 23(1–2).:40–47.

Črešnar, B. & Petrič, Š. (2011). Cytochrome P450 enzymes in the fungal kingdom. In *Biochimica et Biophysica Acta - Proteins and Proteomics*, 1814(1)., 29–35 Elsevier. 10.1016/j.bbapap.2010.06.020

Da Luz, J. M. R., Paes, S. A., Ribeiro, K. V. G., Mendes, I. R. & Kasuya, M. C. M. (2015). Degradation of green polyethylene by Pleurotus ostreatus. *PLoS ONE*, 10(6). 10.1371/journal.pone.0126047

Daly, P., Cai, F., Kubicek, C., Jiang, S., Grujic, M., Rahimi, M., Sheteiwy, M., Giles, R., Riaz, A., de Vries, R., Akcapinar, G., Wei, L. & Druzhinina, I. (2021). From lignocellulose to plastics: Knowledge transfer on the degradation approaches by fungi. *Biotechnology Advances*, 50, 107770.

Danso, D., Chow, J. & Streita, W. R. (2019). Plastics: Environmental and biotechnological perspectives on microbial degradation. In *Applied and Environmental Microbiology* 85, Issue 19 10.1128/AEM.01095-19

Das, N. & Chandran, P. (2011). Microbial degradation of petroleum hydrocarbon contaminants: An overview. *Biotechnology Research International*, 2011. 10.4061/2011/941810

Das, M. & A. Adholeya. (2021). Role of microorganisms in remediation of contaminated soil, in T. Satyanarayana and B. N. Johri, eds. *Microorganisms in Environmental Management: Microbes and Environment*, Dordrecht: Springer Netherlands, pp. 81–111 da Silva, F.J.G. & Gouveia, R.M., 2020. *Cleaner production tools and environmental management practices. Cleaner Production.* Cham: Springer, pp. 153–245

Della Ventura, B., Rea, I., Caliò, A., Giardina, P., Gravagnuolo, A., Funari, R., Altucci, C., Velotta, R. & De Stefano, L. (2016). Vmh2 hydrophobin layer entraps glucose: A quantitative characterization by label-free optical and gravimetric methods. *Applied Surface Science,* 364, 201–207.

Enfrin, M., Hachemi, C., Hodgson, P. D., Jegatheesan, V., Vrouwenvelder, J., Callahan, D. L., Lee, J. & Dumée, L. F. (2021). Nano/micro plastics – Challenges on quantification and remediation: A review. *Journal of Water Process Engineering,* 42, 102128. 10.1016/j.jwpe.2021.102128

Environmental Protection Agency (EPA). (2017). Scope, fate, risks and impacts of microplastic pollution in irish freshwater systems. Report no. 210. Available at: https://www.epa.ie/pubs/reports/research/water/RR%20210Essentra_web.pdf

Fang, G., Tang, B., Liu, Z., Gou, J., Zhang, Y., Xu, H. & Tang, X. (2014). Novel hydrophobin-coated docetaxel nanoparticles for intravenous delivery: In vitro characteristics and in vivo performance. *European Journal of Pharmaceutical Sciences,* 60, 1–9

Fakhrul, H., Fazal, A., Farooq, R., Sohaib, R., Abdul, G. & Muhammad, S. (2014). Assessment of biodegradability of PVC containing cellulose by white rot fungus. *Malaysian Journal of Microbiology.* 10.21161/mjm.55113

Falkenstein, P., Gräsing, D., Bielytskyi, P., Zimmermann, W., Matysik, J., Wei, R. & Song, C. (2020). UV pretreatment impairs the enzymatic degradation of polyethylene terephthalate. *Frontiers in Microbiology,* 11. 10.3389/fmicb.2020.00689

Filiatrault-Chastel, C., Heiss-Blanquet, S., Margeot, A. & Berrin, J. (2021). From fungal secretomes to enzymes cocktails: The path forward to bioeconomy. *Biotechnology Advances,* 52, 107833.

G7 Environment Ministers' Meeting (2019). Available online: https://www.elysee.fr/admin/upload/default/0001/04/7d84becef82b656c246fa1b26519567ce3755600.pdf

G7 Summit (2015). Leaders' Declaration, G7 Summit, 7–8 June 2015. Available online: https://sustainabledevelopment.un.org/content/documents/7320LEADERS%20STATEMENT_FINAL_ CLEAN.pdf

Gallo, F., Fossi, C., Weber, R., Santillo, D., Sousa, J., Ingram, I., Nadal, A. & Romano, D. (2018). Marine litter plastics and microplastics and their toxic chemicals components: the need for urgent preventive measures. In *Environmental Sciences Europe* 30(1)., 10.11 86/s12302-018-0139-z

Gaur, N., Narasimhulu, K. & Y, P. (2018). Recent advances in the bioremediation of persistent organic pollutants and its effect on environment. *Journal of Cleaner Production*, 198, 1602–1631.

GESAMP (2016). "Sources, fate and effects of microplastics in the marine environment: Part two of a global assessment" (Kershaw, P.J., and Rochman, C.M., eds). (IMO/FAO/UNESCO-IOC/ UNIDO/WMO/IAEA/UN/ UNEP/UNDP Joint Group of Experts on the Scientific Aspects of Marine Environmental Protection). Rep. Stud. GESAMP No. 93, 220 p.

Giacomucci, L., Raddadi, N., Soccio, M., Lotti, N. & Fava, F. (2019). Polyvinyl chloride biodegradation by Pseudomonas citronellolis and Bacillus flexus. *New Biotechnology*, 52. 10.1016/j.nbt.2019.04.005

Glazer AN & Nikaido H. (1995). *Microbial biotechnology: fundamentals of applied microbiology*. New York: Freeman

Guan, N., Li, J., Shin, H. dong, Du, G., Chen, J. & Liu, L. (2017). Microbial response to environmental stresses: from fundamental mechanisms to practical applications. In *Applied Microbiology and Biotechnology* 101(10). 10.1007/s00253-017-8264-y

Haider, T., Völker, C., Kramm, J., Landfester, K. & Wurm, F. (2018). Plastics of the future? The impact of biodegradable polymers on the environment and on society. *Angewandte Chemie International Edition*, 58(1)., 50–62.

Hamidian, A. H., Ozumchelouei, E. J., Feizi, F., Wu, C., Zhang, Y. & Yang, M. (2021). A review on the characteristics of microplastics in wastewater treatment plants: A source for toxic chemicals. In *Journal of Cleaner Production* 295, 10.1016/j.jclepro.2021.126480

Hawksworth, D. L. (1991). The fungal dimension of biodiversity: Magnitude, significance, and conservation. *Mycological Research*, 95 (6)., 641–655.

Heckman, D. S., Geiser, D. M., Eidell, B. R., Stauffer, R. L., Kardos, N. L. & Hedges, S. B. (2001). Molecular evidence for the early colonization of land by fungi and plants. *Science*, 293 (5532)., 1129–1133.

Hernández-Arenas, R., Beltrán-Sanahuja, A., Navarro-Quirant, P. & Sanz-Lazaro, C. (2021). The effect of sewage sludge containing microplastics on growth and fruit development of tomato plants. *Environmental Pollution*, 268. 10.1016/j.envpol.2020.115779

Hibbett, D. S., Binder, M., Bischoff, J. F., Blackwell, M., Cannon, P. F., Eriksson, O. E., Huhndorf, S., James, T., Kirk, P. M., Lücking, R., Thorsten Lumbsch, H., Lutzoni, F., Matheny, P. B., Mclaughlin, D. J., Powell, M. J., Redhead, S., Schoch, C. L., Spatafora, J. W., Stalpers, J. A., Vilgalys, R., Aime, M. C., Aptroot, A., Bauer, R., Begerow, D., Benny, G. L., Castlebury, L. A., Crous, P. W., Dai, Y.-C., Gams, W., Geiser, D. M., Griffith, G. W., Gueidan, C., Hawksworth, D. L., Hestmark, G., Hosaka, K., Humber, R. A., Hyde, K. D., Ironside, J. E., Kõljalg, U., Kurtzman, C. P., Larsson, K.-H., Lichtwardt, R., Longcore, J., Miądlikowska, J., Miller, A., Moncalvo, J.-M., Mozley-Standridge, S., Oberwinkler, F., Parmasto, E., Reeb, V., Rogers, J. D., Roux, C., Ryvarden, L., Sampaio, J. P., Schüßler, A., Sugiyama, J., Thorn, R. G., Tibell, L., Untereiner, W. A., Walker, C., Wang, Z., Weir, A., Weiss, M., White, M. M., Winka, K., Yao, Y.-J. & Zhang, N. (2007). 'A higher-level phylogenetic classification of the Fungi'. *Mycological Research*, 111 (5)., 509–547.

Huntington, A., Corcoran, P. L., Jantunen, L., Thaysen, C., Bernstein, S., Stern, G. A. & Rochman, C. M. (2020). A first assessment of microplastics and other anthropogenic particles in Hudson Bay and the surrounding eastern Canadian Arctic waters of Nunavut. *Facets*, 5(1)., 432–454. 10.1139/FACETS-2019-0042

Hwang, J.H., Sadmani, A., Lee, S.J., Kim, K.T., Lee, W.H. (2020). Microalgae: an eco-friendly tool for the treatment of wastewaters for environmental safety. in Bharagava, R.N., Saxena, G. Eds. *Bioremediation of Industrial Waste for Environmental Safety.* Singapore: Springer, pp. 283–304

Janssen, M., Van Leeuwen, M., Van Kooten, T., De Vries, J., Dijkhuizen, L. & Wösten, H. (2004). Promotion of fibroblast activity by coating with hydrophobins in the β-sheet end state. *Biomaterials*, 25(14)., 2731–2739.

Jeon, H., Durairaj, P., Lee, D., Ahsan, M. M. & Yun, H. (2016). Improved NADPH regeneration for fungal cytochrome P450 monooxygenase by co-expressing bacterial glucose dehydrogenase in resting-cell biotransformation of recombinant yeast. *Journal of Microbiology and Biotechnology*, 26(12). 10.4014/jmb.1605.05090

Jenkins, S., Quer, A.M.I., Fonseca, C., Varrone, C. (2019). Microbial degradation of plastics: New plastic degraders, mixed cultures and engineering strategies. in Jamil, N., Kumar, P., and Batool, R. Eds. *Soil Microenvironment for Bioremediation and Polymer Production*, pp. 213–238. 10.1002/9781119592129

Kane, I. A. & Clare, M. A. (2019). Dispersion, accumulation, and the ultimate fate of microplastics in deep-marine environments: A review and future directions. In *Frontiers in Earth Science* (Vol. 7)., 10.3389/feart.2019.00080

Karamanlioglu, M., Houlden, A. & Robson, G. D. (2014). Isolation and characterisation of fungal communities associated with degradation and growth on the surface of poly(lactic) acid (PLA) in soil and compost. *International Biodeterioration and Biodegradation*, 95(PB). 10.1016/j.ibiod.2014.09.006

Katyal, D., Kong, E. & Villanueva, J. (2020). Microplastics in the environment: impact on human health and future mitigation strategies. *Environmental Health Review*, 63(1)., 10.5864/d2020-005

Kawai, F., Kawabata, T. & Oda, M. (2019). Current knowledge on enzymatic PET degradation and its possible application to waste stream management and other fields. In *Applied Microbiology and Biotechnology* 103(11)., 10.1007/s00253-019-09717-y

Kettner, M. T., Rojas-Jimenez, K., Oberbeckmann, S., Labrenz, M. & Grossart, H. P. (2017). Microplastics alter composition of fungal communities in aquatic ecosystems. *Environmental Microbiology*, 19(11). 10.1111/1462-2920.13891

Khan, S., Nadir, S., Shah, Z. U., Shah, A. A., Karunarathna, S. C., Xu, J., Khan, A., Munir, S. & Hasan, F. (2017). Biodegradation of polyester polyurethane by *Aspergillus tubingensis*. *Environmental Pollution*, 225. 10.1016/j.envpol.2017.03.012

Kim, D.Y. & Rhee, Y.H. (2003). Biodegradation of microbial and synthetic polyesters by fungi. *Applied Microbiology and Biotechonology*, 61, 300–308.

Krueger, M. C., Hofmann, U., Moeder, M. & Schlosser, D. (2015). Potential of wood-rotting fungi to attack polystyrene sulfonate and its depolymerisation by *Gloeophyllum trabeum* via hydroquinone-driven Fenton chemistry. *PLoS ONE*, 10(7). 10.1371/journal.pone.0131773

Kumar, N. & Min, K. (2011). Phenolic compounds biosorption onto *Schizophyllum commune* fungus: FTIR analysis, kinetics and adsorption isotherms modeling. *Chemical Engineering Journal*, 168(2)., 562–571.

Lee, H., Shim, W. J. & Kwon, J. H. (2014). Sorption capacity of plastic debris for hydrophobic organic chemicals. *Science of the Total Environment*, 470–471, 10.1016/j.scitotenv.2013.08.023

Liu, J., He, J., Xue, R., Xu, B., Qian, X., Xin, F., Blank, L., Zhou, J., Wei, R., Dong, W. & Jiang, M. (2021). Biodegradation and up-cycling of polyurethanes: Progress, challenges, and prospects. *Biotechnology Advances*, 48, 107730.

Loredo-Treviño, A., Gutiérrez-Sánchez, G., Rodríguez-Herrera, R. & Aguilar, C. N. (2012). Microbial enzymes involved in polyurethane biodegradation: A review. In *Journal of Polymers and the Environment* 20(1)., 10.1007/s10924-011-0390-5

Lu, T., Zhang, Q.L. & Yao, S.J. (2016). Application of biosorption and biodegradation functions of fungi in wastewater and sludge treatment. in Purchase D. eds. Fungal Applications in Sustainable Environmental Biotechnology. *Fungal Biology*. Cham: Springer. 10.1007/978-3-319-42852-9_4

Magnin, A., Pollet, E., Perrin, R., Ullmann, C., Persillon, C., Phalip, V. & Avérous, L. (2019). Enzymatic recycling of thermoplastic polyurethanes: Synergistic effect of an esterase and an amidase and recovery of building blocks. *Waste Management*, 85, 141–150.

Magnin, A., Pollet, E., Phalip, V. & Avérous, L. (2020). Evaluation of biological degradation of polyurethanes. *Biotechnology Advances*, 39, 107457.

McLean, K., Sabri, M., Marshall, K., Lawson, R., Lewis, D., Clift, D., Balding, P., Dunford, A., Warman, A., McVey, J., Quinn, A., Sutcliffe, M., Scrutton, N. & Munro, A. (2005). Biodiversity of cytochrome P450 redox systems. *Biochemical Society Transactions*, 33(4)., 796–801.

Mohanan, N., Montazer, Z., Sharma, P. & Levin, D. (2020). Microbial and enzymatic degradation of synthetic plastics. *Frontiers in Microbiology*, 11.

Muenmee, S., Chiemchaisri, W. & Chiemchaisri, C. (2016). Enhancement of biodegradation of plastic wastes via methane oxidation in semi-aerobic landfill. *International Biodeterioration and Biodegradation*, 113. 10.1016/j.ibiod.2016.03.016

Muhonja, C. N., Makonde, H., Magoma, G. & Imbuga, M. (2018). Biodegradability of polyethylene by bacteria and fungi from Dandora dumpsite Nairobi-Kenya. *PLoS ONE*, 13(7). 10.1371/journal.pone.0198446

Newsham, K. K. (2012). Fungi in extreme environments. In *Fungal Ecology* 5(4). 10.1016/j.funeco.2012.04.003

Olicón-Hernández, D. R., González-López, J. & Aranda, E. (2017). Overview on the biochemical potential of filamentous fungi to degrade pharmaceutical compounds. In *Frontiers in Microbiology* 8, Issue SEP. 10.3389/fmicb.2017.01792

Oliveira, J., Belchior, A., da Silva, V. D., Rotter, A., Petrovski, Ž., Almeida, P. L., Lourenço, N. D. & Gaudêncio, S. P. (2020). Marine environmental plastic pollution: Mitigation by microorganism degradation and recycling valorization. In *Frontiers in Marine Science* (Vol. 7). 10.3389/fmars.2020.567126

Paço, A., Duarte, K., da Costa, J. P., Santos, P. S. M., Pereira, R., Pereira, M. E., Freitas, A. C., Duarte, A. C. & Rocha-Santos, T. A. P. (2017). Biodegradation of polyethylene microplastics by the marine fungus Zalerion maritimum. *Science of the Total Environment*, 586. 10.1016/j.scitotenv.2017.02.017

Palm, G. J., Reisky, L., Böttcher, D., Müller, H., Michels, E. A. P., Walczak, M. C., Berndt, L., Weiss, M. S., Bornscheuer, U. T. & Weber, G. (2019). Structure of the plastic-degrading Ideonella sakaiensis MHETase bound to a substrate. *Nature Communications*, 10(1). 10.1038/s41467-019-09326-3

Pathak, V. & Navneet. (2017). Review on the current status of polymer degradation: A microbial approach. *Bioresources and Bioprocessing*, 4(1).

Piscitelli, A., Cicatiello, P., Gravagnuolo, A. M., Sorrentino, I., Pezzella, C. & Giardina, P. (2017). Applications of functional amyloids from fungi: Surface modification by class I hydrophobins. In *Biomolecules* 7(3)., 10.3390/biom7030045

Quintella, C., Mata, A. & Lima, L. (2019). Overview of bioremediation with technology assessment and emphasis on fungal bioremediation of oil contaminated soils. *Journal of Environmental Management*, 241, 156–166.

Rana, K. I. (2019). Usage of potential micro-organisms for degradation of plastics. *Open Journal of Environmental Biology*. 10.17352/ojeb.000010

Rochman, C. M., Kross, S. M., Armstrong, J. B., Bogan, M. T., Darling, E. S., Green, S. J., Smyth, A. R. & Veríssimo, D. (2015). Scientific evidence supports a ban on microbeads. In *Environmental Science and Technology* 49(18)., 10.1021/acs.est.5b03909

Russell, J. R., Huang, J., Anand, P., Kucera, K., Sandoval, A. G., Dantzler, K. W., Hickman, D. S., Jee, J., Kimovec, F. M., Koppstein, D., Marks, D. H., Mittermiller, P. A., Núñez, S. J., Santiago, M., Townes, M. A., Vishnevetsky, M., Williams, N. E., Vargas, M. P. N., Boulanger, L. A., Strobel, S. A. (2011). Biodegradation of polyester polyurethane by endophytic fungi. *Applied and Environmental Microbiology*, 77(17). 10.1128/AEM.00521-11

Rydz, J., Sikorska, W., Kyulavska, M. & Christova, D. (2014). Polyester-Based (bio)degradable polymers as environmentally friendly materials for sustainable development. *International Journal of Molecular Sciences*, 16(1)., 564–596.

Sánchez, C. (2020). Fungal potential for the degradation of petroleum-based polymers: An overview of macro- and microplastics bio-degradation. *Biotechnology Advances*, 40, 107501.

Sangeetha Devi, R., Rajesh Kannan, V., Nivas, D., Kannan, K., Chandru, S. & Robert Antony, A. (2015). Biodegradation of HDPE by Aspergillus spp. from marine ecosystem of Gulf of Mannar, India. *Marine Pollution Bulletin*, 96(1–2)., 32–40. 10.101 6/j.marpolbul.2015.05.050

Sankhla, I. S., Sharma, G. & Tak, A. (2020). Fungal degradation of bioplastics: An overview. In *New and Future Developments in Microbial Biotechnology and Bioengineering*. 10.1016/b978-0-12-821007-9.00004-8

Schwartz, M., Perrot, T., Aubert, E., Dumarçay, S., Favier, F., Gérardin, P., Morel-Rouhier, M., Mulliert, G., Saiag, F., Didierjean, C. & Gelhaye, E. (2018). Molecular recognition of wood polyphenols by phase II detoxification enzymes of the white rot Trametes versicolor. *Scientific Reports*, 8(1). 10.1038/s41598-018-26601-3

Shin, J., Kim, J., Lee, Y. & Son, H. (2018). Fungal cytochrome P450s and the P450 complement (CYPome) of *Fusarium graminearum*. *Toxins*, 10(3)., 112.

Smith, M., Love, D. C., Rochman, C. M. & Neff, R. A. (2018). Microplastics in seafood and the implications for human health. In *Current environmental health reports* 5(3)., 10.1007/s40572-018-0206-z

Sowmya, H. V., Ramalingappa, Krishnappa, M. & Thippeswamy, B. (2014). Degradation of polyethylene by Trichoderma harzianum—SEM, FTIR, and NMR analyses. *Environmental Monitoring and Assessment*, 186(10). 10.1007/s10661-014-3875-6

Spina, F., Tummino, M., Poli, A., Prigione, V., Ilieva, V., Cocconcelli, P., Puglisi, E., Bracco, P., Zanetti, M. & Varese, G. (2021). Low density polyethylene degradation by filamentous fungi. *Environmental Pollution*, 274, 116548.

Tahir, L., Ishtiaq Ali, M., Zia, M., Atiq, N., Hasan, F. & Ahmed, S. (2013). Production and characterization of esterase in *Lantinus tigrinus* for degradation of polystyrene. *Polish Journal of Microbiology*, 62(1)., 101–108.

Tao, J., Wang, Y., Xiao, Y., Yao, P., Chen, C., Zhang, D., Pang, W., Yang, H., Sun, D., Wang, Z., et al. (2017). One-step exfoliation and functionalization of graphene by hydrophobin for high performance water molecular sensing. *Carbon*, 116, 695–702.

Tedersoo, L., Sánchez-Ramírez, S., Kõljalg, U., Bahram, M., Döring, M., Schigel, D., May, T., Ryberg, M. & Abarenkov, K. (2018). 'High-level classification of the Fungi and a tool for evolutionary ecological analyses', *Fungal Diversity*, 90 (1)., 135–159.

Thompson, R. C., Olson, Y., Mitchell, R. P., Davis, A., Rowland, S. J., John, A. W. G., McGonigle, D. & Russell, A. E. (2004). Lost at sea: Where is all the plastic? *Science*, 304(5672)., 10.1126/science.1094559

Thompson, R. C., Moore, C. J., Saal, F. S. V. & Swan, S. H. (2009). Plastics, the environment and human health: Current consensus and future trends. In *Philosophical Transactions of the Royal Society B: Biological Sciences* 364(1526)., 10.1098/rstb.2009.0053

Urbanek, A. K., Rymowicz, W. & Mirończuk, A. M. (2018). Degradation of plastics and plastic degrading bacteria in cold marine habitats. *Applied Microbiology and Biotechnology*, 102 (18)., 7669–7678.

Velimirovic, M., Tirez, K., Voorspoels, S. & Vanhaecke, F. (2021). Recent developments in mass spectrometry for the characterization of micro- and nanoscale plastic debris in the environment. *Analytical and Bioanalytical Chemistry*, 413(1)., 10.1007/s00216-020-02898-w

Velzeboer, I., Kwadijk, C. J. A. F. & Koelmans, A. A. (2014). Strong sorption of PCBs to nanoplastics, microplastics, carbon nanotubes, and fullerenes. *Environmental Science and Technology*, 48(9)., 10.1021/es405721v

Viegas, C., Pinheiro, A.C., Sabino, R., Viegas, S., Brandao, J. & Verissimo, C. (2016). *Environmental mycology in public health – fungi and mycotoxins risk assessment and management academic press.* https://doi.org/10.1016/C2012-0-03608-3

Volke-Seplveda, T., Saucedo-Castaeda, G., Gutirrez-Rojas, M., Manzur, A. & Favela-Torres, E. (2002). Thermally treated low density polyethylene biodegradation by Penicillium pinophilum and Aspergillus niger. *Journal of Applied Polymer Science*, 83(2). 10.1 002/app.2245

Wang, J., Liu, X., Li, Y., Powell, T., Wang, X., Wang, G. & Zhang, P. (2019). Microplastics as contaminants in the soil environment: A mini-review. In *Science of the Total Environment* 691, 10.1016/j.scitotenv.2019.07.209

Walker, G.M. & White, N.A. (2017). Introduction to fungal physiology. *Fungi: biology and applications*, pp. 1–35.

Wessels, J.G.H. (1997). Hydrophobins: Proteins that change the nature of the fungal surface in Poole, R. ed. *Advances in Microbial Physiology, Volume 38*. Elsevier Science & Technology.

Woodall, L. C., Sanchez-Vidal, A., Canals, M., Paterson, G. L. J., Coppock, R., Sleight, V., Calafat, A., Rogers, A. D., Narayanaswamy, B. E. & Thompson, R. C. (2014). The deep sea is a major sink for microplastic debris. *Royal Society Open Science*, 1(4). 10.1098/rsos.140317

Woodall, L. C., Sanchez-Vidal, A., Paterson, G. L. J., Coppock, R., Sleight, V., Calafat, A., Rogers, A. D., Narayanaswamy, B. E. & Thompson, R. C. (2014). The deep sea is a major sink for microplastic debris Subject category: Subject Areas: Author for correspondence: *Royal Society Open Science*, 1(4).

World Health Organization. (2019). Microplastics in drinking-water. Licence: CC BY-NC-SA 3.0 IGO.

Wösten, H. A. B. & Scholtmeijer, K. (2015). Applications of hydrophobins: current state and perspectives. In *Applied Microbiology and Biotechnology* 99(4). 10.1007/s00253-014-6319-x

Wright, S. L. & Kelly, F. J. (2017). Plastic and human health: A micro issue? *Environmental Science and Technology*, 51(12)., 10.1021/acs.est.7b00423

Wu, Y., Li, J., Yang, H. & Shin, H. (2017). Fungal and mushroom hydrophobins: A review. *Journal of Mushroom*, 15(1)., 1–7.

Yamada-Onodera, K., Mukumoto, H., Katsuyaya, Y., Saiganji, A. & Tani, Y. (2001). Degradation of polyethylene by a fungus, *Penicillium simplicissimum* YK. *Polymer Degradation and Stability*, 72(2). 10.1016/S0141-3910(01)00027-1

Yousif, E. & Haddad, R. (2013). Photodegradation and photostabilization of polymers, especially polystyrene: Review. In*SpringerPlus* 2(1). 10.1186/2193-1801-2-398

Yuan, J., Ma, J., Sun, Y., Zhou, T., Zhao, Y. & Yu, F. (2020). Microbial degradation and other environmental aspects of microplastics/plastics. *Science of the Total Environment*, 715. 10.1016/j.scitotenv.2020.136968

Zhang, S., Gedalanga, P.B., Mahendra, S. (2017). Advances in bioremediation of 1,4-di-oxane-contaminated waters. *J. Environ. Manag.* 204, 765–774.

Zhang, S., Merino, N., Okamoto, A. & Gedalanga, P. (2018). 'Interkingdom microbial consortia mechanisms to guide biotechnological applications'. *Microbial biotechnology*, 11 (5)., 833–847.

Index

Note: Page numbers in *Italic* refer to figures; and in **Bold** refer to tables.